Keeping the Farm
In Your Family

Keeping the Farm In Your Family

What You *Need* to Know

MARK R. ALVIG & DAVID P. BUSS
Contributing Editor GARY OTTE

ADAMS
BUSINESS &
PROFESSIONAL

KEEPING THE FARM IN YOUR FAMILY: What You Need to Know.
© copyright 2012 by Mark R. Alvig and David P. Buss. All rights reserved. No part of this book may be reproduced in any form whatsoever, by photography or xerography or by any other means, by broadcast or transmission, by translation into any kind of language, nor by recording electronically or otherwise, without permission in writing from the author, except by a reviewer, who may quote brief passages in critical articles or reviews.

Estate planning is unique to each individual. This guide is intended as a starting point to provide general information. It is not intended to give any specific legal or tax advice. Every plan needs careful, personal analysis. When developing an estate plan, consult with a team of advisors, including your Attorney, CPA, Accountant, Financial Planner, Investment Advisor, Life Insurance Professional, and Property/Casualty Professional.

ISBN 10: 1-59298-225-5
ISBN 13: 978-1-59298-225-7

Library of Congress Catalog Number: 2011941920

Printed in the United States of America

First Printing: 2012

16 15 14 13 12   5 4 3 2 1

Cover and interior design by James Monroe Design, LLC.

Adams Business & Professional
A Division of Beaver's Pond Press, Inc.
7104 Ohms Lane, Suite 101
Edina, MN 55439–2129
(952) 829-8818
www.BeaversPondPress.com

To order, visit www.BeaversPondBooks.com
or call (800) 901-3480. Reseller discounts available.

CONTENTS

Introduction . . . . *v*

1 Estate Planning: What It Is and Why It Matters to You . . . 1.1

2 Your Estate Taxes and the Taxpayer Relief Act of 1997 . . . . 2.1

3 The 2010 Tax Relief Act 3.1

4 Wills and Other Ways 4.1

5 Planning For Liquidity 5.1

6 How to Use Trusts in Estate Planning 6.1

7 Planning: "Just In Case…" 7.1

8 Getting More Out of Giving—Charities and Your Estate . . 8.1

9 The Family-Owned Business 9.1

10 Issues Farm Families Face 10.1

11 Buy-Sell Agreements 11.1

12 Medical Assistance—What You Need to Know 12.1

Glossary of Estate Planning Terms *13.1*

INTRODUCTION

Historically, estate planning has been perceived as a resource for the wealthy.

Today, wealth is commonplace. Estate planning has become significantly more than writing a will. The aging of our society and increased life expectancies create more issues than just disposition of assets. Health, health care, asset preservation, business transfer, estate tax reduction, and probate avoidance are all issues of estate planning.

Estate planning is the most challenging area of financial planning. This book contains a list of ten common mistakes in estate planning because the majority of people who have started estate plans have made mistakes. Mistakes may cause families to have countless problems handling an estate. Some mistakes may cost heirs dearly.

The purpose of this book is to be a general guide designed to cover a broad range of essential estate planning topics. It is in a format that demystifies the topics and brings significant awareness of a wide range of issues that must be dealt with for effective, comprehensive estate planning.

Accomplishing an effective, well-thought-out estate plan is a challenging process. The techniques examined in *Keeping the Farm in Your Family* will help in making correct choices when beginning or continuing an estate planning journey.

1

Estate Planning:
What It Is and Why It Matters to You

"Estate planning." For most people, that term seems intimidating. That's why so many people put off doing something that would make life easier for themselves and for their loved ones.

But estate planning really doesn't need to be intimidating. After all, since you plan for so many other things in life, it just makes good sense to put a little time and thought into this aspect of your life. It's well worth it.

What is an *estate*? In simple terms, it's everything that you own.

In English common law, the basis for our legal system, the term "estate" meant the land to which a person was lawfully entitled. The

> **Estate:** All your assets, including your home, real estate, bank accounts, securities, retirement plans, and insurance policies.

Old English term meant material condition or status. Today the term refers to anything that you own now and anything that you might own in the future.

Estate planning involves all aspects of your life. It involves thinking about what can happen to you, to your health, and to your family. That means that you need to consider such matters as your assets and your debts, your needs and your interests as you grow older, and how you can best take care of yourself and the people close to you.

If you own a business, estate planning also involves ensuring that your business will be in good hands after you retire or when you die.

If you care about your family and everything that you've worked to acquire and to build, then estate planning should be important to you. There's a lot you need to do to ensure an independent future for yourself and a generous endowment for your loved ones.

Estate planning requires time and effort. What's involved?

- You need to make sure that you know what you own. This means taking inventory of your property and other assets.
- You need to know the facts about your ownership, to determine and document legal ownership.
- You may also need to retitle (change legal ownership), transfer, or sell certain assets.
- You need to identify beneficiaries and heirs and to determine how much and how soon they will receive their share of your estate, perhaps by setting up trusts.
- You need to make decisions about how to reduce the impact of estate taxes and how to fund them to keep the government from unnecessarily shrinking your estate.
- You need to prepare for health care.

These Are the Major Areas Involved in Estate Planning.

Some people—far too many—just say, "It will all work out." That's a comforting thought.

Unfortunately, far too often they're wrong. And they pay, financially and emotionally, for believing and trusting in fate. And their loved ones pay as well.

Many people believe the members of their family are capable of handling just about anything. They trust that their loved ones will be able to divide up estate assets fairly and with little difficulty. They think the family bond is so solid that the many details of distributing property, securities, cash, other assets, and personal items will just happen in an orderly fashion without any planning.

The reality is that all too often it doesn't work out that way. Depending on the legal complexity of the situation and the value of the assets, an estate can be held up in probate for years. That means that the courts have to sort everything out—and maybe make some decisions. This situation can strain even the most secure family relationships.

The simple fact is that many families have suffered because somebody they loved failed to consider what might happen and neglected to prepare for the future. Old age and death cause enough suffering. Why cause further sorrow and increase the emotional and financial burdens?

> **Retitle:** To change the legal ownership (title) of property.
>
> **Beneficiary:** Person who receives benefits from such assets as trusts, insurance policies, and estates.
>
> **Heir:** Person who inherits property when somebody dies.

> **Probate:** Court and legal proceedings that settle all the legal and financial matters for somebody who dies owning property.

With effective and thoughtful planning, you can minimize the burden on your loved ones by protecting and directing the distribution of your assets. You can decide how you want to take care of your family and your possessions with minimal or no involvement of the probate court.

You can reduce or eliminate estate and income taxes, avoid many administration requirements, and facilitate the difficult process of settling your estate. If you plan wisely, you can avoid the many problems that too often leave family members feuding or confused, perhaps even deprived of what would make their lives easier.

Through estate planning, you can take care of the people you love and the things you own. It's your family. They're your things. The decision is up to you. With this book to guide you, you can do whatever you think best to prepare for the future when you will no longer be able to take care of what matters most to you.

2

Your Estate Taxes and the Taxpayer Relief Act of 1997

Uncle Sam doesn't really care whether you're dead or alive—at least in terms of when you give away your property. Your gifts (what you give away while living) or bequests (what you give away when you die) are lumped together as subject to estate and gift taxes.

The estate and gift tax is a transfer tax assessed on property that changes possession. It's called a *unified* tax because the same tax rates, deductions, and rules apply to both gifts and estate.

Uncle Sam has a heart when it comes to generosity. The tax laws establish a unified credit exemption. What this means is that you can give away a certain amount of assets, during your lifetime or upon your death, without incurring a federal estate or gift tax. That concept seems simple enough, but it results in some complicated calculations.

> **Transfer Tax:** A tax imposed when ownership of property passes from one person to another.

A Little History

To understand the impact of the Taxpayer Relief Act of 1997, let's get some recent historical perspective of the gift and estate tax laws in the United States.

In 1981, the federal estate tax laws provided an exemption of $175,625. In other words, if an adjusted gross estate was less than that amount, it paid no estate tax. In 1982 the law was changed. Under the revised law, the exemption began to rise until 1987, when it reached $600,000. That was the magic figure for ten years: If the adjusted gross estate was less than $600,000, no federal estate taxes were imposed.

> **Unified Credit Exemption:** Amount of property that an individual can give away through gifts or an estate.

The Taxpayer Relief Act of 1997 provided for the estate tax exemption to again increase incrementally, beginning in 1998 with a $625,000 unified credit exemption and ending in 2006 at $1,000,000. If your generosity (gifts and estate) exceeds that amount, it's subject to marginal tax rates, starting at 37% and reaching 60%, then dropping to 55%. (See **Figure 2A**.)

> **Marginal Tax:** The tax imposed on an estate that is valued in excess of the unified credit exemption, when the value of lifetime gifts has been included.

In addition, Uncle Sam recognizes that families that own businesses should not be treated the same as other families. The Taxpayer Relief Act of 1997 provides special estate tax relief for family-owned businesses. If a business meets certain criteria, it will be eligible for an additional exemption. There are restrictions on this special exemption, which we'll examine later in this chapter.

Figure 2A: Estate Tax Exemptions under the Taxpayer Relief Act of 1997

Year	Exemption Amount (a)	Unified Credit (b)
1999	$650,000	$211,300
2000	$675,000	$220,550
2001	$675,000	$220,550
2002	$700,000	$229,800
2003	$700,000	$229,800
2004	$850,000	$287,300
2005	$950,000	$326,300
2006	$1,000,000	$345,800

Gift Taxes

Under the old law, an individual could not give anyone any gifts that exceeded $3,000 per person per year without creating a gift tax liability for the generous individual. But that limit did not take into account a very important economic reality—inflation.

The Taxpayer Relief Act of 1997 provides for inflation. The new law indexes the maximum amount allowable $10,000, starting in 1999. The method for indexing is based on the Consumer Price Index (CPI), which is a federal measurement of the value of a dollar, in terms of what we can buy with our money.

> **Gift Tax:** A tax imposed on the transfer of property while alive, to be paid by the person giving the gift, if the value of the gift(s) is greater than the annual allowable limit.

That's good news, in theory. However, because of how the indexing is calculated—in increments of $1,000—that max may remain at $13,000 for a while. In fact, given the current low rate

> **Consumer Price Index (CPI):** A federal measurement of inflation and deflation based on changes in the relative costs of goods and services for a typical consumer.

2.3

of inflation, it may be at least a few years before we see any substantial increase.

If you're married, you and your spouse can combine your maximum amounts. You can do this even if the property belongs to just one of the spouses—even if you're not living in one of the community property states. That's nice of Uncle Sam—or maybe it just makes sense, since spouses can give each other gifts without incurring any gift taxes.

What's a gift? Generally when people talk about "gifts," it seems fairly straightforward. A gift is something of some value that you give to somebody. But when the IRS gets involved, we need a definition and, of course, some exceptions.

> **Gift:** Any voluntary transfer of property or property interests to another without adequate consideration.

So, a gift is any transaction in which property or property interests are voluntarily transferred to another without adequate consideration. This definition does not include future interests, gifts that the recipient can use only in the future.

Here are a few exceptions. The federal gift tax does NOT apply to:

- Gifts to spouse
- Gifts to tax-exempt organizations, such as charities and religious entities
- Gifts to schools
- Gifts to governmental agencies
- Loans to family members at interest rates below the market rate
- College tuition paid directly to the institution (not the student!)
- Medical expenses paid directly to the institution (not the patient!)

Note: If you make over $13,000 in gifts to a qualified charitable organization, you must file IRS Form 709, the Federal Gift Tax Return. You won't pay a gift tax, but the IRS likes to keep track of big gifts. The $13,000 is indexed to the CPI.

YOUR ESTATE TAXES AND THE TAXPAYER RELIEF ACT OF 1997

What happens if you exceed the exemption amount for gifts to one person in a single year? Then you must file IRS Form 709. Of course, you don't need to put yourself in that situation, if you do the proper financial planning.

Estate Taxes

How does the estate tax work? First, an estate is inventoried to determine its total value. There are certain deductions that can be subtracted from the estate, including funeral expenses, administration and attorney's fees, income taxes to be paid, and debts and mortgages.

> **Estate Tax:** A tax imposed on the estate when it transfers to heirs (also known as inheritance tax or death tax).

The adjusted value of the estate is compiled. That amount is then referenced on the unified federal estate and gift tax table. (See **Figure 2B**.) From the estate tax, the estate is allowed to deduct the unified credit to arrive at the taxes that are due. The unified credit amount is equal to the estate taxes on an equivalent estate exemption.

Figure 2B: Unified Federal Gift and Estate Tax Table

Dollar Value of Taxable Transfer (a)	Federal Unified Tax Before Credits (b)	Percent of Excess (c)
$0	$0	18%
$10,000	$1,800	20%
$20,000	$3,800	22%
$40,000	$8,200	24%
$60,000	$13,000	26%
$80,000	$18,200	28%
$100,000	$23,800	30%
$150,000	$38,800	32%
$200,000	$54,800	32%
$250,000	$70,800	34%
$300,000	$87,800	34%
$400,000	$121,800	34%
$500,000	$155,800	37%

Dollar Value of Taxable Transfer (a)	Federal Unified Tax Before Credits (b)	Percent of Excess (c)
$600,000	$192,800	37%
$700,000	$229,800	37%
$750,000	$248,300	39%
$800,000	$267,800	39%
$900,000	$306,800	39%
$1,000,000	$345,800	41%
$1,100,000	$386,800	41%
$1,250,000	$448,300	43%
$1,500,000	$555,800	45%
$1,600,000	$600,800	45%
$2,000,000	$780,800	49%
$2,100,000	$829,800	49%
$2,500,000	$1,025,800	53%
$2,600,000	$1,078,800	53%
$3,000,000	$1,290,800	55%
$10,000,000	$5,140,800	60%
$21,225,000	$11,875,800	55%

Key:
(a) Base value of taxable estate. (b) Base tax on estate.
(c) Percent tax on excess.

Rate drops to 55% over $21,225,000

The terminology makes this calculation seem more complicated than it actually is. For example, in 2005 the unified credit is $555,800, which is the amount of tax due on a $1,500,000 estate. The significance of this method of calculation is that the estate does not deduct the credit exemption first, but rather after the tax level is determined. Thus, the estate is taxed at the highest possible level before the exemption is used.

THE BOTTOM LINE

The increase in unified credits and corresponding equivalent exemptions enacted in the Taxpayer Relief Act of 1997 looks good—at first glance. A closer examination reveals that the $600,000 exemption that was available in 1987, then indexed at a 3% annual rate, would have been $806,350 by year 1997. In year 2006 it projects to $1,052,104. (See **Figure 2C**.) Many estates are growing at a much faster rate than inflation.

Conclusion: The new estate tax law is not quite keeping pace with inflation. Estate planning is more important than ever.

Figure 2C: Indexing of the Unified Credit from 1987 to 2006

Year	Projected Increase Annually (a)	Projected Increase Exemption (b)	Existing and Future Exemption (c)	Net Difference Projected (d)
1987	$0	$600,000	$600,000	$0
1988	$18,000	$618,000	$600,000	($18,000)
1989	$18,540	$636,540	$600,000	($36,540)
1990	$19,096	$655,636	$600,000	($55,636)
1991	$19,669	$675,305	$600,000	($75,305)
1992	$20,259	$695,564	$600,000	($95,564)
1993	$20,867	$716,431	$600,000	($116,431)
1994	$21,493	$737,924	$600,000	($137,924)
1995	$22,138	$760,062	$600,000	($160,062)
1996	$22,802	$782,864	$600,000	($182,864)
1997	$23,486	$806,350	$600,000	($206,350)
1998	$24,190	$830,540	$625,000	($205,540)
1999	$24,916	$855,457	$650,000	($205,457)
2000	$25,664	$881,120	$675,000	($206,120)
2001	$26,434	$907,554	$675,000	($232,554)
2002	$27,227	$934,780	$700,000	($234,780)
2003	$28,043	$962,824	$700,000	($262,824)
2004	$28,885	$991,709	$850,000	($141,709)
2005	$29,751	$1,021,460	$950,000	($71,460)
2006	$30,644	$1,052,104	$1,000,000	($52,104)

Key:
(a) Hypothetical projected annual increase of estate tax exemption indexed from 1987 at 3%. (b) Hypothetical projected estate tax exemption indexed from 1987 at 3%. (c) Actual estate tax exemption through year 2006. (d) Hypothetical projected shortfall of actual exemption compared to indexed 3% exemption since 1987.

Unlimited Marital Deduction

Under the Federal Uniform Estate and Gift Tax rules, there are provisions that, if used properly, allow estates to reduce the impact of estate taxes. One of these rules is the unlimited marital deduction. Simply put, a person can leave an estate of any size to his or her surviving spouse without paying any estate taxes.

If the estates of both spouses combined are over $1,000,000, you should do some planning so that each of you can take advantage of your own $1,000,000 exemption. If you don't use your exemption, it disappears when you die: Your surviving spouse can't use it.

> **Unlimited Marital Deduction:** A legal provision that allows a person to give gifts of any size or pass an estate of any size to his or her spouse, without any gift or estate taxes or using any part of the unified credit.

(Why do we use an exemption of $1,000,000 here if this exemption is scheduled to rise almost every year until 2010? We'll answer that question with another: When are you going to die? Death is just a fact of life. We intend to live for years, but we must plan as if we could die at any moment. That's why we use the 2006 exemption in our explanation.)

If, however, the combined estates were considerably higher, over $2,000,000, then you'd have to pay some taxes if you were to die at this moment. If that seems unlikely, you'll want to continue reading. We've got some suggestions for managing that estate so the tax burden will be less.

How to Reduce Estate Taxes

Using the unlimited marital deduction will certainly reduce taxes when the first spouse dies.

However, the estate that passes to the surviving spouse increases his or her estate. That growth is ultimately taxed, when the second spouse dies. So using the marital deduction can hurt your heirs. To avoid this unfortunate tax situation, it's wise to balance the value of both spouse's estates whenever possible.

This estate planning point is too often overlooked. This occurs because the estate tax system is a marginal tax rate system. It taxes an estate at the top brackets of the adjusted estate's value. The tax is imposed on the value of an estate in excess of the unified credit.

Figure 2D illustrates a scenario from 2006. (Again, we're using current figures for the sake of simplicity. We could show the same principles at work for 2006, with the exemption and credit that will be in effect that year.)

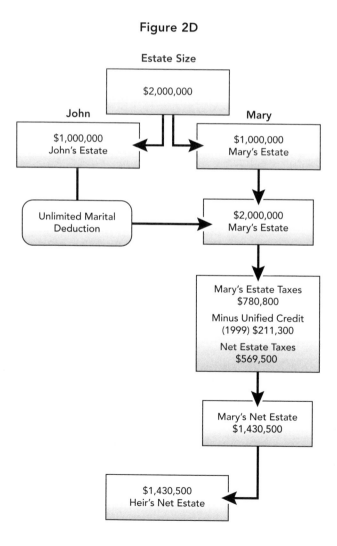

Figure 2D

Figure 2E

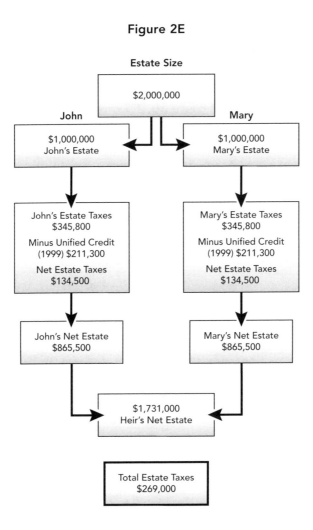

John and Mary have an estate valued at $2,000,000. John dies, leaving all of his estate to Mary. Because of the unlimited marital deduction, Mary receives John's $1,000,000 without any loss to estate taxes. When Mary dies, however, the entire $2,000,000 is subject to estate taxes.

The gift and federal estate tax table (**Figure 2B**) shows those taxes to be $780,800. With the current (2006) unified credit of $211,300 (**Figure 2A**, the unified credit equivalent to the $650,000 exemption), the taxes amount to $569,500. So Uncle Sam takes a 28.475% bite out of the estate.

YOUR ESTATE TAXES AND THE TAXPAYER RELIEF ACT OF 1997

Much of that $569,500 loss for John's and Mary's heir was unnecessary, since John's unified credit estate exemption of $650,000 went unused.

Now, let's try a little estate planning. **Figure 2E** shows how we can structure their estate to reduce that tax bite.

When John dies, his $1,000,000 goes to his heirs, not to Mary. The taxes on his estate are $345,800 (**Figure 2B**), but with his unified credit estate exemption of $211,300 (**Figure 2A**, the unified credit equivalent to the $650,000 exemption), the federal estate tax drops to $134,500. The estate tax liability on the combined estates of John and Mary is reduced by $269,000.

That's much better than the $569,500 lost in the scenario in **Figure 2D**. That simple strategy of willing $1,000,000 directly to the heir reduced the tax bite to 13.45%.

Note this very important and interesting point: The amount of estate tax saved here by distributing John's and Mary's estates separately exceeded the value of John's unified credit that went unused when his estate passed to Mary. His unified credit was $211,300, but when used properly it saved $300,500 in estate taxes.

This estate planning point is too often overlooked. This occurs because the estate tax system is a marginal tax rate system. It taxes an estate at the top brackets of the adjusted estate's value.

The tax is imposed on the value of an estate in excess of the unified credit.

John and Mary used a simple strategy to keep a lot of money away from Uncle Sam for their heir. We'll discuss more advanced estate planning techniques in later chapters. We will also look at using the marital deduction in combination with other strategies.

THE GENERATION-SKIPPING TAX

Tax planners have often recommended trusts and other devices as ways to pass property down through several generations with no estate tax burden. The advantage was to provide income from the property to the following generation without incurring any estate transfer tax, since the heirs would not inherit any rights to the property itself. In other words, the transfer would skip a generation or more.

The Tax Reform Act of 1976 imposed a new tax and new tax return requirement on such an arrangement. This Act was repealed and replaced by legislation in the Tax Reform Act of 1986. The generation-skipping transfer tax is imposed on every generation-skipping transfer, whether in trust or through direct distributions. However, the generation-skipping transfer tax does not apply to lifetime annual exclusion gifts to individuals and to certain trusts or to certain transfers for medical or educational purposes.

> **Generation-Skipping Tax:** Tax imposed on trusts and other devices that would otherwise allow you to provide your heirs with income from property yet bypass the estate transfer tax.

Each person has a $3,500,000 (2009) exemption from the generation-skipping tax. The exemption has been indexed to the consumer price index formula. However, the law that created the increases to the GST Tax Exemption is set to expire in 2011. This means that the GST exemption amount will return to $1,000,000 per individual in 2011 unless Congress makes additional changes.

Disclaimers

Here's another simple way to reduce or avoid estate taxes. And it's not an idea that would occur to most people.

We're all familiar with the term "disclaimer" as it's generally used. It's a way to not take responsibility for something—usually through a lot of fine print. In the case of estate planning, the term has a somewhat different meaning: What the person is disclaiming is property.

Let's put that difference into legal terms. A disclaimer is an irrevocable and unqualified refusal by a person to accept an interest in property. The disclaimed property will be treated as though it had never been transferred to the person making the disclaimer. This action can suit various positive purposes and provide significant benefits, financial and otherwise.

Why would anybody want to refuse to accept taking possession of inherited property? As we said, it's not an idea that would occur to most people. There are various situations in which it might make sense to disclaim an inheritance. Consider the following examples:

- A named beneficiary can essentially make a tax-free gift to another person by disclaiming an inheritance—if the inheritance would, under state law, go to the second person.

> **Disclaimer:** Refusal to accept property that is passed on through a will.

- A will leaves the entire estate to two children, but unequally. The child designated to receive the larger share might disclaim part of that share in order to make the split more equitable.

- A surviving spouse doesn't need the property. He or she might disclaim it in order to avoid estate taxes when his or her property passes on later to the children.

- A child who is financially comfortable inherits a share of a parent's estate. He or she could disclaim that inheritance so that it passes into a trust established in the will for his or her children, to add to the share designated specifically for them.

- A significant portion of a taxable estate is held in the husband's name and the wife's estate is less than the personal exemption amount. The wife dies first; in her will she leaves everything to the husband. The husband disclaims the assets, which then pass to the heirs, using a portion of the deceased spouse's unified credit/personal exemption.

Disclaiming the inheritance can be an effective strategy when a will is outdated. Let's take an example.

When your rich uncle drew up his will twenty years ago, you (his favorite nephew) were struggling to pay some bills and take care of your two young children. So he willed all his $100,000 to you, with the provision that it would go to your kids if you died before your uncle. Unfortunately, your uncle dies without updating his will to take into account the current situation. Now, twenty years later, you're quite wealthy and your two kids are trying to make it through graduate school.

You know that your uncle would want his money to go to your children now, since they need it more than you do. Of course, you could accept the inheritance and then give the money to your children. But then you'd have to give them only $11,000 each per year or pay gift taxes. Either way, they lose. A disclaimer allows you to do what your uncle would want. By using a disclaimer, you can, in effect, step out of the way of the inheritance and let it go directly to your children.

A disclaimer can also be useful when there's no will at all. If the state laws dictate that the inheritance should pass to you, rather than to another relative, who would be second in line to inherit, you can respect the unwritten wishes of the deceased and disclaim the inheritance, so that it passes to the next heir in line.

So, how do you make a disclaimer?

For the purposes of federal estate, gift, and generation-skipping transfer taxes, you must meet the following requirements:

- The refusal must be in writing.
- The disclaimer must be irrevocable and unqualified.
- The written refusal must be received by the transferor, his or her legal representative, or the holder of the legal title not more than nine months after the date on which the transfer is made or the date on which that person reaches age 21, whichever comes later.
- The person making the disclaimer must not have accepted the property or any of its benefits.
- The property must pass to someone other than the person making the disclaimer.
- The property must pass without any direction from the person disclaiming the property. That means he or she cannot direct to whom the disclaimed property should go. (Before deciding to disclaim, the person should know who is next in line to receive the property, according to the terms of the will and/or state law.)

You've got to be careful about meeting each of these requirements. If you don't, you will be considered the legal recipient of the property. But that's not all: You will then be deemed to have gifted the property that you were trying to disclaim. In other words, you may achieve the outcome you desired—the beneficiary may receive the property—but achieving it through a gift transfer rather than through a proper disclaimer can mean you'll be paying a gift tax for your generosity. Also, if the recipient of your generosity is two generations below you, the generation-skipping transfer tax may also be imposed. Ouch!

We should mention that you don't have to be generous to an extreme. It's possible to make only a partial disclaimer, that is, to disclaim only part of the property that you would otherwise inherit. In this case as well, you must meet all the statutory requirements listed above.

In addition, a partial disclaimer of property will be respected only on the following two conditions:

- The portion disclaimed must be severable from the part accepted.
- The disclaimer must identify the specific assets being disclaimed.

In other words, you cannot disclaim half of the family mansion (the house cannot be severed) or half of the sports car collection (the description is too vague).

As you might guess, despite the extra complications, this means of accepting some property while disclaiming the rest can be quite useful.

Any beneficiary considering disclaiming an inheritance should consult with an estate planner. If the property disclaimed is large enough, it could result in a generation-skipping transfer tax. Some cynics claim that no good deed goes unpunished. If you're not careful, you could prove that point.

Changes for Family-Owned Businesses

As we mentioned earlier, when it comes to estate taxes, Uncle Sam treats a family differently if it owns a business. There's particular relief in the Taxpayer Relief Act of 1997 for family-owned businesses.

The new law created an estate tax exemption that can mean a total exemption of as much as $1.3 million for qualified family-owned businesses. This estate tax exemption can be taken only to the extent that it exceeds the personal estate exemption. In other words, when the unified credit exempts $625,000 of an estate, the business exemption covers an additional $675,000.

That's a great benefit for a family-owned business. However, it will be a little less good as the years pass. That $1,300,000 remains constant but, as we have discussed, the personal estate exemption rose incrementally to $1,000,000 in 2006. As a result, the business benefit declined to a $300,000 equivalent exemption by 2006. (See **Figure 2F**.)

Figure 2F: Estate Tax Exemptions under the Tax Relief Act of 1997

Year	Family Exemption Amount (a)	Maximum Business Exclusion (b)	Combined Exclusion (c)
1999	$650,000	$650,000	$1,300,000
2000	$675,000	$625,000	$1,300,000
2001	$675,000	$625,000	$1,300,000
2002	$700,000	$600,000	$1,300,000
2003	$700,000	$600,000	$1,300,000
2004	$850,000	$450,000	$1,300,000
2005	$950,000	$350,000	$1,300,000
2006	$1,000,000	$300,000	$1,300,000
Key:			
(a) The amount each individual estate is allowed as an exemption. (b) The new family business exclusion. The business will have restrictions for eligibility. (c) The maximum combined exclusion for a family-owned business cannot exceed $1,300,000.			

So much for the math. Now it's time for some legal definitions.

Who can take advantage of that new exemption? To qualify, family-owned businesses must meet the following requirements:

1. Any interest in a trade or business (regardless of the actual form of the business)
2. With a principal place of business in the United States
3. If the ownership is held at least 50% by one family, 70% by two families, or 90% by three families, and
4. As long as the decedent's family owns at least 30% of the trade or business.

How does the law define "family"? To qualify for inclusion in the above percentages, you must be in one of the following categories: The decedent's spouse, the decedent's parents, the decedent's grandparents, the decedent's children and their spouses, the decedent's spouse's children and their spouses, and the decedent's brothers and sisters and their spouses.

The total value of the qualified family-owned business interest that is passed to qualified heirs must be more than 50% of the decedent's adjusted gross estate. This is known as the 50% test. It's more complicated than we're presenting it here, so again we advise consulting an estate planner.

Now, we've got another term to define—qualified heirs. To qualify, individuals must be members of the decedent's family (as defined above) and meet both of the following qualifications:

1. They must also have been actively employed by the trade or business for at least ten years prior to the date of death.
2. They must continue to materially participate in the trade or business for at least five years out of any eight-year period within ten years following the decedent's death.

Failure to comply with this second point will cause a scheduled pro rata recapture of the appropriate tax from all qualified heirs, not just the one in violation of the provisions. What that means, in plain English, is that if any heir doesn't keep working in the family business, it will cost the other members of the family. The IRS knows how to split heirs!

There are additional provisions and points to keep in mind. Here's one more: The business will fail to qualify for this special tax treatment if more than 35% of the adjusted ordinary gross income of the business for the year of the decedent's death was personal holding company income.

The details would fill another chapter. If this provision of the Taxpayer Relief Act of 1997 seems to apply to you, an estate planner can go through the details and help you get some additional relief.

3

THE 2010 TAX RELIEF ACT

OVERVIEW

On June 7, 2001, President Bush signed into law the Economic Growth and Tax Relief Reconciliation Act of 2001 (EGTRRA). Among the many changes to the tax law, also know as the Bush-era Tax Cuts, was estate tax repeal effective for the year 2010. To avoid the 2010 estate tax repeal, the House of Representatives voted to permanently extend the 2009 system ($3.5 million exemption, 45% tax rate) but this system failed to pass in the Senate.

On December 17, 2010, President Obama signed into law the Tax Relief, Unemployment Insurance Reauthorization, and Job Creation Act of 2010 (the 2010 Tax Relief Act). This act extends all of the Bush-era Tax Cuts through 2012 and contains significant changes in many areas including:

- Estate Tax
- Gift Tax

- Generation-Skipping Transfer Tax
- Capital Gains and Tax Rate Dividends
- Income Tax

The 2010 Tax Relief Act reunifies the estate and gift tax exemptions for 2011 and 2012. This permits taxpayers to transfer up to $5 million per person, and $10 million per couple, tax-free during life. In addition, it imposes a maximum rate of 35% for all three types of transfer tax. The 2012 exemption amounts are indexed for inflation in increments of $10,000.

This act is scheduled to sunset on December 31, 2012. This provides a narrow window for use of the enhanced $5 million exemptions and low 35% rates as illustrated below. In 2013, the estate, and gift tax exemptions are scheduled to return to $1 million.

	2010		2011—2012		2013	
TAX	Exemption	Rate	Exemption	Rate	Exemption	Rate
Gift	$1M	35%	$5M	35%	$1M (?)	55% (?)
Estate	$5M	35%	$5M	35%	$1M (?)	55% (?)
GST	$5M	0%	$5M	35%	$1M (?)	55% (?)

Estate Tax Opportunities

The effective date of the estate tax exemption was January 1, 2010. The estates of 2010 decedents are taxed at a 35% estate tax rate with a stepped-up basis and a $5 million exemption. All appreciated assets transferred at death in 2011 and 2012 will also receive a step-up in basis. Property with a stepped-up basis receives a basis equal to the property's fair market value on the date of the decedent's death (or on an alternate valuation date).

Although the 2010 Tax Relief Act imposes an estate tax on decedents dying in 2011 and 2012 at a rate of 35%, tax will not be due unless the taxable estate exceeds the decedent's available estate tax exemption.

Executors of individuals dying in 2010 can elect to use the $5 million estate tax exemption and step-up in basis, apply the prior law that allows 2010 decedents to be subject to no estate taxes, with a modified carryover basis for estate assets.

Under a modified carryover basis, the executor may increase the basis of estate property only by a total of $1.3 million, with other estate property taking a carryover basis equal to the lesser of the decedent's basis or the fair market value of the property on the decedent's death.

The specific circumstances of each estate will determine which set of laws is more beneficial to apply.

Portability of Unused Estate Tax Exemption

Previously, to take full advantage of a husband's or wife's estate tax exemption, the first spouse's exemption amount must be transferred at death either to beneficiaries other than the surviving spouse, or be held in a "credit shelter" or "bypass" trust. This requires sophisticated estate planning and it also requires married couples to unnaturally divide the ownership of their assets. Assets transferred to these trusts at the death of the first spouse are "sheltered" from estate taxation at the death of the second spouse.

The 2010 Tax Relief Act provides for "portability" between spouses of the estate tax applicable exclusion amount. This allows the surviving spouse to "elect" to use any unused exemption of the first spouse to die, providing the surviving spouse with a larger exclusion amount. This act also eliminates the ability to accumulate exclusion amounts from serial marriages. This provision is only available in 2011 and 2012 and requires that an estate tax return be filed at the first spouse's death.

While this provision will simplify estate planning for married couples, it should be noted that there are still benefits to using a "credit shelter" or "bypass" trust that the portability exemption does not offer. These benefits include creditor protection, protection from a later divorce, and keeping the increased value of trust assets out of the second spouse's taxable estate.

Extension of Time to File Returns

Because some of the transfer tax changes are retroactive to January 1, 2010, the act extends the due date for any estate and generation-skipping tax return for the estates of decedents dying, and generation-skipping transfers made during 2010 until September 17, 2011.

Gift Tax Opportunities

The new $5 million per person gift tax exemption presents an opportunity for substantial tax-free gifting in 2011 and 2012. This exemption has never been this high and is scheduled to return to $1 million in 2013.

Generation-Skipping Transfer Tax Opportunities

Unlike the gift tax or estate tax, the generation-skipping transfer (GST) tax can apply either during life or at death. For generation-skipping transfers, the 2010 Tax Relief Act allows individuals to make aggregate transfers of up to $5 million to "skip persons" outright or in trust tax-free. "Skip persons" include family members two or more generations younger than the transferor, as well as non-family members more than thirty-seven-and-a-half years younger than the transferor.

Any transfers made in excess of the $5 million exemption amount will be subject to the 35% GST tax. For generation-skipping transfers in 2010, the act retroactively imposes a 0% tax and a $5 million exemption amount. The $5 million may be used to exempt gifts to trusts that are expected to benefit multiple generations, so that generation-skipping transfers from the trusts in subsequent years are also exempt from GST. This provision is set to expire at the end of 2012.

CAPITAL GAINS AND DIVIDEND TAX RATES

This act extends the 15% maximum rate for capital gains and qualified dividends. In 2013, the maximum long-term capital gains rate is scheduled to be 20%.

INCOME TAX OPPORTUNITIES

The 2010 Tax Relief Act gives all businesses the ability to write off 100% of the cost of certain machinery and equipment placed into service from September 9, 2010 through December 31, 2011. In 2012, the write-off is reduced to 50% of the cost. Other important income and employment tax provisions include the following:

- Income tax brackets. The provision extends the 10%, 15%, 25%, 28%, 33%, and 35% individual income tax rates until 2013, with the rate structure indexed for inflation. In 2013, ordinary income is scheduled to be taxed at a maximum rate of 39.6%. And net investment income will likely be subject to a new 3.8% surtax.

- Personal exemptions and itemized deductions. The act extends the repeal of the phase-outs of personal exemptions and itemized deductions.

- Alternative minimum tax (AMT) "patch." This is a special tax system with a minimum 26% rate, originally designed to ensure that the ultra-affluent "paid their share." The 2010 Tax Relief Act provides a two-year extension of AMT relief—indexing the AMT exemption for inflation for 2010 and 2011.

AMT Exemption	2010	2011
Unmarried Individuals	$47,450	$48,450
Married Surviving Spouses—Joint Filing	$72,450	$74,450
Married Spouses—Separate Filing	$36,225	$37,225

- Individual retirement account (IRA) "charitable rollover" Provision. This provision allows income tax-free transfers directly from an IRA to a qualified charity of up to $100,000 per taxpayer, per taxable year.
- Employee payroll tax. A payroll/self-employment tax holiday during 2011 grants a temporary reduction in the employee side of payroll tax, from 6.2% to 4.2% on wages. The self-employed will pay 10.4% on self-employment income up to the $106,800 ceiling.
- Unemployment insurance. The unemployment insurance proposal provides a one-year reauthorization of Federal unemployment insurance benefits.
- Other extended provisions. Numerous other provisions have been extended including "bonus depreciation," the deductions for college tuition, student loan interest, state and local sales taxes, and the child and adoption tax credits.

Review Estate Plans

To identify and exploit opportunities, as well as identify and mitigate potential risk, consider a careful review of your current estate plan.

State Estate Tax

Life insurance is exempt from estate and inheritance taxes in most states. However, if the insurance benefit goes to the estate or the executor rather than a third party like a spouse or child, it is not exempted. Also, remember that many states still impose an estate tax. In fact, twenty states and the District of Columbia still have an estate tax, an inheritance tax, or both.

Estate tax is based on the overall size of the estate. A top rate in the teens is most common. Minnesota, however, exacts a much higher top rate than the national average with a 41% rate on estates over $1 million.

State	2011 Exemption	2011 Top Rate	State	2011 Exemption	2011 Top Rate
Connecticut	$3,500,000	12	New Jersey	$675,000	16
Delaware	$5,000,000	16	New York	$1,000,000	16
District of Columbia	$1,000,000	16	North Carolina	$5,000,000	16
Hawaii	$3,600,000	16	Ohio	$338,333	7
Maine	$1,000,000	16	Oregon	$1,000,000	16
Maryland	$1,000,000	16	Rhode Island	$850,000	16
Massachusetts	$1,000,000	16	Vermont	$2,750,000	16
Minnesota	$1,000,000	41	Washington	$2,000,000	19

State Inheritance Tax

Beneficiaries are responsible for inheritance taxes. The top rate for inheritance taxes is also generally in the teens. In this category, Indiana has the highest rate at 20 percent on inheritances over $100. The upshot? Residents in some states are much more affected by state taxes than federal taxes.

State	2011 Exemption	2011 Top Rate	State	2011 Exemption	2011 Top Rate
Indiana	$100	20	Nebraska	$10,000	18
Iowa	$0	15	New Jersey	$0	16
Kentucky	$500	16	Pennsylvania	$0	15
Maryland	$150	10	Tennessee	$1,000,000	9.5

4

Wills and Other Ways

There's an old axiom: *Where there's a will, there's a way.* With a twist, that's true legally: A will is a way to dispose of your property. But it's not the only way. In this chapter we'll cover the essentials of a will and explore some ways to keep your assets out of probate.

Wills

A will is to estate planning as notations are to music. It is but one of a complex combination of elements that, skillfully orchestrated, create a harmonious estate plan. Many people know that a will is an integral part of estate planning. They are correct, but planning involves much more.

Very few people really understand the entire scope of a will, how it is administered, and how it influences the distribution of estate assets after death. We'll give a brief

> **Testator/Testatrix:** The man or woman who makes the will and whose estate is to be distributed.

overview of wills, discuss the essentials of a will, and provide an outline for a standard will.

By law, a "last will and testament" is a written document that directs how the testator's property is to be distributed when he or she dies. When that happens, the will is admitted to the probate court and established as a properly executed and valid will. The probate estate assets are then distributed. That's how it works, in general.

Essentials of a Will

We usually think of a will as a document hammered out by an attorney and containing a lot of dry legal language punctuated by the frequent "inasmuch as" and "aforementioned" and "heretofore" and "notwithstanding." But a will can be valid even if not drawn up in this way. You'll want to check the laws of your particular state in this matter, as in all matters of estate planning.

There are certain essentials to keep in mind when making a will that can stand up as valid:

- You must be of legal age. This means at least 18, but the age differs by state.

- You must be of sound mind and memory. What does that mean? That you understand what you're doing and that you know the general nature and extent of your property. That's known as "testamentary capacity." The law presumes mental competence; it's tough for anyone to challenge this point, if you make sure not to neglect any major considerations and to at least mention the expected beneficiaries.

- You must put your will in writing. There are exceptions, but play it safe. It's best to have it typed or printed by a computer.

- You should identify yourself by your legal name.

- You must include at least one substantive provision. That means that your will must actually dispose of your property and indicate that you intend the document to be a will.

- You must date your will. You should also specify where you're signing the will.

- You must sign the document, voluntarily. There are legal ways to get around this point, if you're unable to sign. It's also a good idea (and may be required by your state) to have some wording at the end to attest that this is your will.

- You must have two people witness your will who are not named as beneficiaries in the will. (Some states require three witnesses.) The witnesses must watch you sign the document and they must know that it's your will, although they don't need to read it.

It's not very expensive to have an attorney draw up your will. A basic will can cost between $250 and $1250. If you belong to a group legal service plan, you may be able to have a will drawn free of charge or at least for far less than the usual rate.

There are also will kits that you can use. Check what requirements your particular state has for wills. You certainly don't want probate to decide that your quick 'n' easy will-in-a-box is not valid. You won't have a second chance to get it right.

All in all, your loved ones and your estate are probably worth investing a little more money to do your will up right. How would you feel if even one beneficiary lost out on what you intended to bequeath, simply because you cut corners on your will?

One final point here: An attorney can make sure that your will is legal and valid, but an estate planner can help you decide if it's good, if it will do what you want it to do—and how it fits with the rest of your planning.

Components of a Will

We can't provide you with a sample will that will meet your specific needs and prove valid in every state, but here's an outline of what's included in a typical will.

Introduction: This section gives the legal name and the residence of the testator, affirms that the testator/testatrix is of sound mind and memory, states that this is his or her last will and testament, and revokes any prior wills and codicils.

> **Residuary Estate:** What remains of an estate after all specific property bequests have been made.

Typical wording: "I, name, residing at address, being of sound mind and memory, declare this to be my last will and testament and I hereby revoke all prior wills and codicils." (We'll cover codicils a little later.)

Liabilities: This section provides instructions for paying outstanding debts, funeral and administrative expenses, and any estate taxes.

Typical wording: "I authorize my executor/executrix to pay my enforceable unsecured debts, medical and funeral expenses, and costs incurred in administering my estate. These payments are to come from my residuary estate.

Specific Devises: This section details any bequests of specific assets. It should identify each beneficiary fully by name and include alternate beneficiaries, in the event that a beneficiary has died or wishes to disclaim a bequest.

Typical wording: "I bequeath all my personal effects and household items, such as jewelry, furniture, clothing, and books, to my relationship, Name. If he/she does not survive me or disclaims this property, I bequeath it to relationship, Name.

Or, "I bequeath my two cars and one truck to my relationship, Name. If he/she does not survive me or disclaims this property, I bequeath it to relationship, Name...."

Residuary Estate: This section names a beneficiary or beneficiaries for what remains of the estate after the specific bequests.

Typical wording: "I bequeath my residuary estate to my <u>relationship, Name</u>. If he/she does not survive me or disclaims this property, I bequeath it to <u>relationship, Name</u>."

Personal Representative: This section names the person who will serve as executor or executrix and be responsible for executing the will. There should be a provision for an alternate.

Typical wording: "I appoint my <u>relationship, Name</u>, as executor/executrix of this will. If he/she is unwilling or unable to serve in this capacity, for any reason, I appoint my <u>relationship</u>, <u>Name</u> as successor executor/ executrix. My executor/executrix shall have all the powers granted to executors under the laws of state to exercise all legal powers as he/she determines to be in the best interests of my estate. I direct that no bond or security of any kind be required of my executor."

> **Devise:** (as a noun) A bequest or gift in a will; (as a verb) to bequeath or give in a will.

Signature: This section contains the signatures of the testator/testatrix and the witnesses, with the date and place of the signing.

Typical wording: "I have signed this last will and testament in the presence of the undersigned witnesses on this <u>day of date</u>, <u>year</u>. <u>(Signature)</u>

"Signed and declared by <u>testator/testatrix</u> to be his/her last will and testament, in our presence, who at his/her request, in his/her presence, and in the presence of each other have signed our names as witnesses."

Depending on your circumstances, you may want to add sections to do any of the following:

- Appoint a guardian for any minor children
- Instruct the executor or executrix to set up an account for the children under the Uniform Gifts to Minors Act or the Uniform Transfers to Minors Act

- Forgive certain debts
- Disinherit one or all of your children

Uniform Gifts to Minors Act/Uniform Transfers to Minors Act: Law that allows an account to be set up for a minor, with an adult designated as custodian of the property for the minor, who is the legal owner of the property and has an unrestricted right to it upon reaching the age of majority.

Final Details

Some wills also include instructions regarding funeral arrangements and burial wishes, but some experts advise against including these matters in a will, since it might not be examined until days after a death, by which time other arrangements may have been made.

Letter of Instructions: A memo that contains such information as the location of your will, the location of other vital documents, and any wishes for your funeral and burial.

You should play it safe and express any funeral or burial wishes in advance to your loved ones or to anyone who might be in a position to make the arrangements. It's wisest to write out the details and make copies for these people.

The best way is in what's called a *letter of instructions*. This is a memo of personal details that you should keep with your will. Send a copy to your executor. Keep one for yourself.

The letter of instructions should include such information as the location of your will, the location of other vital documents, and any wishes for your funeral and burial. It costs very little to make sure that people know what you want. It will also help them make decisions at a time that will be very difficult for them emotionally.

So, once you've got a will, where do you put it? The usual reaction is to treat it as you treat all of your important documents and put it in a safety deposit box. If you do that, remember that the bank might seal the box when you die.

When that happens, the box can be opened only when a release is obtained. So keep several copies where they'll be accessible: The person whom you've named as executor of your will should have a copy, of course, as well as your attorney and members of your family.

CHANGES, CHANGES, CHANGES...

We all know that there's nothing certain in life but change. In fact, as soon as you make out your will, the odds are really good that something in your life will change. You may acquire more assets or get rid of some. People may enter your life whom you want to include among your beneficiaries—e.g., a grandchild is born or a child marries. You may lose a beneficiary—e.g., somebody dies, a charitable organization shuts down, or you decide to disinherit a relative or a former friend.

> **Codicil:** A legal change to a will, written and properly witnessed.

So, then, is it back to the will drawing board? Not necessarily.

You can amend your will by adding a codicil. You simply specify the change that you'd like to make, following the same legal procedures as for your will: Put it in writing, date it, and sign it with witnesses.

Because of the formalities involved in making valid codicils and because it's so easy to prepare wills now, it may be better simply to prepare a new will. You'd certainly be wise to do so if the change is major, such as disinheriting a child or changing executors.

One final word here: Ademption. That's a legal term from a Latin word meaning "taking away."

This is when you take away something from your estate, disposing of property that you bequeathed in your will, so that the bequest is invalidated.

> **Ademption:** The removal of property from an estate by the owner after he or she has bequeathed it in a will.

Here's an example. You draw up your will and leave to your favorite niece that 1955 Rolls-Royce Silver Dawn that she's loved since she was two. But then, a few years later, you run into a sudden liquidity problem, so you sell the Silver Dawn. Three months after that, you come into a little money, so you immediately go out and buy a 1959 Rolls-Royce Phantom V. After all, your niece probably won't know the difference. Ah, but she will—because that Phantom won't go to your niece, since it's not the vehicle that you specified in your will.

Our advice here is to anticipate changes, then prepare for them. If you provide in your will to leave a specific piece of property, you might want to include a provision that would cover the possibility of selling or exchanging the property.

INTESTATE: THE SLOW ROUTE THROUGH PROBATE

The term "intestate" describes a person who dies without leaving a will. A person who dies without a will creates the very real potential for assets to be distributed in ways that he or she never intended. If you don't express your wishes in a will, you're leaving it to state law to write your will for you. For your loved ones, that may be a most unfortunate, time-consuming, and expensive mistake.

> **Intestate:** The state of having died without leaving a will.

Intestacy creates the worst possible situation: The distribution of every piece of probate property in the estate must be decided through the probate process. The best situation, in contrast, would be to avoid probate altogether. In recent times there has been an abundance of information distributed on various methods of doing this. We'll discuss these methods later in this chapter.

You understand why it's a very bad idea to leave to probate all the decisions about your estate.

But why do people want to avoid the probate process entirely? The most common incentives for avoiding probate are that the process takes time, costs money, and is a public record subject to anyone's inspection at the county courthouse. We will now consider these issues in more detail.

Probate

What is probate? The word comes from the Latin, *probare*, meaning "to prove." It's a legal procedure, supervised by the court, to prove a will is valid. That's essentially all that probate needs to be—assuming that there's a will. If not, as we've stressed, probate makes up for that missing document of closure.

That's the concept of probate. But, of course, things are always more complicated in practice.

> **Probate:** The legal procedure to determine, if there's a will, whether the will is valid, or if there's no will, how the estate should be distributed.

The Probate Process: Usual Scenarios

The probate process gathers all the assets in the estate, takes out whatever is needed to pay debts and taxes and administrative expenses, and then distributes the assets among the beneficiaries. That process can take time, cost a little or a lot, and open up your life to anybody with an interest or merely idle curiosity. Let's take a look:

Time. Generally, the probate process will take from four months to two years. It can take even longer. How long it takes to settle an estate is determined by the following factors:

- Whether there's a valid will
- What types of assets make up the estate
- How well the estate's assets are organized
- Whether there are trusts
- Whether the probate process is formal or informal
- How competent the executor is
- How many beneficiaries there are
- How easily the beneficiaries can be located

When it's necessary to liquidate property, that process takes time. This applies particularly to real estate, business assets, or unique property (unlike publicly traded stock). Another factor that can affect how long probate takes is challenges. Wills can also be challenged by neglected potential heirs or dissatisfied beneficiaries, who may disagree with distribution or even raise the question of testamentary capacity.

> **Cost.** The cost of probate is difficult to predetermine, as it is affected by many of the same influences mentioned above. The probate laws for the particular state you are in will also factor into the probate cost.
>
> Probate costs rise as complications multiply. Some states have unique statutory probate processes that tend to inflate costs.
>
> **Publicity.** Probate court documents are public records. Anyone who wants to research a probate estate's records can simply go to the courthouse for that information.

Does that matter to you? Maybe. Maybe not. But it's something to keep in mind if there's any information about your estate that you'd like to keep confidential. If ongoing business or personal issues warrant, you should consider strategies that will keep some or all of your estate out of probate, to ensure privacy.

That's a quick description of how probate works for people who die testate, that is, with a will. It gets a whole lot worse if there's no will.

> **Testate:** The state of having died with a will.

The Probate Process: Worst-Case Scenarios

When a person dies intestate, the probate process must also, in essence, make a will after the fact. Here again, what happens depends to some extent on the particular state.

State law will determine what happens to your property. This process is called "intestate succession." Your property will be distributed to your spouse and children or, if you have neither, to other relatives according to a statutory formula. If no relatives can be found to inherit your property, it will go into your state's coffers. Also, in the absence of a will, a court will determine who will care for your young children and their property if the other parent is unavailable or unfit.

You can consult the probate code for your state to find out the line of intestate succession. You can also consult with an attorney. But first, you should answer this question: Why put any effort into knowing how a probate court is likely to distribute your property if you can take action to keep that important matter out of the hands of strangers?

> **Intestate Succession:** The process for determining what will happen to the property and any minor children of a person who dies without a will.

Do we dare emphasize yet again that dying without a will creates a scenario in which your assets may be distributed in ways that you never intended? We hope that, at this point, you are determined to make out a will—and interested in exploring ways to make sure that at least some of your property will bypass the probate process.

BYPASSING PROBATE

There are several ways to keep property from passing through probate. First, we'll discuss how certain types of assets and certain types of titling allow us to bypass probate entirely.

As we noted earlier, a will distributes assets within a person's probate estate. Many times testators expect that their wills are going to take care of distributing all their assets, but they would later roll over in their graves as their wills fail to do as expected. Why? Because of the types of assets or how they are owned.

TYPES OF ASSETS

Certain types of assets by their very nature rarely end up being subject to probate, as we will discuss. Asset types fall into several main categories:

- Real estate
- Stocks and bonds
- Mutual funds
- Bank accounts
- Annuities
- Life insurance
- Individual Retirement Accounts (IRAs)
- 401(k) plans
- Profit-sharing plans
- Pensions
- Miscellaneous tax-qualified retirement plans
- Personal property

The type of asset will determine what titling choices are available, that is, the ways in which an asset may be owned. How an asset is titled will determine whether an asset can be distributed without a will and thus avoid probate.

Titling: Ownership of an asset, usually as individual owner, joint tenants, or tenants in common.

Figure 4A shows how the asset classes listed above can be owned and whether or not the assets will be included among the probate assets.

Figure 4A: Type of Ownership Generally Available and If Typically Subject To Probate

	Individual Ownership	Tenants in Common	Joint Tenancy	Probate Avoidance Available*	Probate Assets**
Real Estate	Yes	Yes	Yes	Yes (4)	Yes
Stocks & Bonds	Yes	Yes	Yes	Yes (5)	Yes
Mutual Funds	Yes	Yes	Yes	Yes (5)	Yes
Bank Accounts	Yes	Yes	Yes	Yes (6)	Yes
Annuities	Yes (1)	Yes (2)	Yes (3)	Yes	No
Life Insurance	Yes (1)	Yes (2)	Yes (3)	Yes	No
IRAs	Yes (1)	No	No	Yes	No
401(k)s	Yes (1)	No	No	Yes	No
Profit Sharing	Yes (1)	No	No	Yes	No
Pensions	Yes (1)	No	No	Yes	No
Personal Property	Yes	No	Yes	Yes (7)	Yes

Key:

(1) Assets typically owned by Individual. Non-probate beneficiary designation available. (2) Non-qualified annuities and life insurance may, in some instances, be held as Tenancy-In-Common. (3) Non-qualified annuities and life insurance may, in some instances, be held in Joint Tenancy. (4) Real estate may be held in life estate, or a quit claim deed can be executed during lifetime. (5) Mutual funds, stocks, bonds, and securities can have transfer on death designations. (6) Bank accounts may have a payable on death or Totten Trust designation. (7) A will can authorize preparations of a list to handle distribution of personal property.

*The ability to name a beneficiary and avoid probate.
**Assets typically subject to probate if owned Individually or as Tenancy-In-Common, or if the "estate" is beneficiary.

TYPES OF OWNERSHIP

Figure 4A shows three types of ownership: Individual ownership, tenancy in common, and joint tenancy. These are the most common methods of titling used. Certain titling of ownership passes assets outside of probate directly to the beneficiary or joint owner. In this section, we will explore ownership titling and how it affects estate distribution.

Individual ownership: This is, as you would expect, when only one person owns the property. Individual ownership generally allows for distribution by will. Examples could include real estate, securities, and bank accounts.

Joint tenancy and joint tenancy with right of survivor: These are the most common types of asset ownership between husband and wife. When one of the joint tenants dies, his or her interest passes to the surviving spouse automatically. This type of titling does not allow for distribution by will. This is only logical, because these two types of ownership are joint. There is no probate requirement.

> **Joint Tenants-in-Common:** Owners of a shared asset, with the interest of any owner, upon death, becoming part of that person's estate.
>
> **Joint Tenants with Rights of Survivorship:** Owners of a shared asset, with the interest of any owner, upon death, passing to the surviving co-owners.

This titling may seem attractive, but it can have adverse consequences for estate planning. However, for many estates it is an appropriate form of titling.

Tenancy in common: This titling allows individuals' undivided interest in assets to be distributed by a will. As with joint tenancy, two or more individuals are on the title of the asset. Although this type of titling is available for many assets, it is most commonly used for real estate.

Titling can be particularly important if you live in a community property state (Arizona, California, Louisiana, Nevada, New Mexico, Texas, Washington, and Wisconsin). The laws in these states provide that most property acquired during marriage is held equally by husband and wife. There are exceptions, notably property acquired by inheritance or as a gift. It's wise to settle possible questions of ownership before the time comes to settle your estate.

Ownership of assets has a significant impact on how well an estate plan will work and what impact the will may have when the estate is settled. Questions of ownership can be quite confusing.

People may ask themselves, "Just what does my will do for me?" The answer: A will distributes only assets that are in your probate estate.

Estate planning for distribution of assets is very involved. It depends a great deal on asset titling.

What about those assets that don't end up in a probate estate? How are they easily recognized? The assets that do not typically end up in a probate estate include life insurance, annuities, retirement plans (such as IRAs, 401(k)s, profit sharing, pensions, and tax-sheltered accounts) and assets owned in joint tenancy. With the exception of joint tenancy, which transfers title to the surviving joint tenant(s), all of these types of assets provide for a beneficiary designation that allows for distribution without going through probate.

LIFE ESTATE

A life estate is an additional type of ownership that avoids probate. It is actually a lifetime transfer, but the person making the transfer, the life tenant, retains all use of the property. In most cases, the life tenant cannot sell or mortgage the property without the permission of the specified new owner, called the remainderman.

A typical use of this titling is parents who transfer real estate to their children while reserving the right to use the property until death. A life estate can also be used by a person to provide a home for his or her grandparents as life tenants. As such, the home would then pass to the person's children when the grandparents die.

> **Life Tenant:** A person who makes a lifetime transfer of a property and retains all use of the property for life.
>
> **Remainderman:** A person who has a future interest in a life estate or a trust.

Since the life estate is an actual lifetime transfer of title, the property is not subject to probate and a will has no impact on it, since the life tenant is no longer the owner. However, the value of the property is included in the estate of the life tenant for estate tax purposes. Life estate is a simple, very effective tool in estate planning, but as with all estate planning strategies, it must be analyzed and carefully and compared against other strategies.

Trusts: Another Way to Avoid Probate

Finally, we conclude our discussion of wills and probate by noting that several types of trusts allow you to bypass probate. However, that's only one advantage of trusts. We'll continue to discuss the various types of trusts and their advantages in Chapter 6, "How to Use Trusts in Estate Planning."

Beneficiary Deeds

Beneficiary deeds closely resemble life estates with the only exception being that the original owner retains all rights. In the event of death, the rights pass on to the named beneficiary without probate.

5

Planning For Liquidity

Let's talk *liquidity*. That's just a fancy word for the ease with which an asset can be turned into cash. Cash is liquid because we can use it for almost every sort of economic need or interest.

(Of course, if you were in the middle of a desert, for example, you might not be able to trade cash for water—a real liquid. But in almost any situation, cash is liquidity to the extreme.)

> **Liquidity:** The degree to which an asset can be converted into cash quickly and without loss of value.

Degrees of Liquidity

We distinguish between liquid assets and non-liquid or illiquid assets, but that's a simplification. Liquidity is usually relative, depending on the situation. If you can sell an asset for a fair price and in little or no time, then that asset is liquid. If, on the other hand, it takes a long time to sell an asset at a fair price, that asset is not very liquid. So, when we use the terms "liquid" and "non-liquid" or "illiquid" or "fixed," we're really discussing areas along a continuum.

Assets that are liquid, besides cash, would range from bank accounts (savings or checking) and certificates of deposit to investments such as savings bonds, Treasury Bills, money market funds, mutual funds, stocks, and municipal bonds or municipal bond mutual funds. Keep in mind that differences exist among liquid assets.

For example, unlike bank accounts, if you suddenly had to redeem your shares in a mutual fund—even a money market mutual fund—your shares might be worth more or less than you paid for them. Assets that are not very liquid would include real estate, motor vehicles, collectibles, personal property, thinly traded securities, and closely held stock (stock in a company owned by family members or a small group of individuals but usually not the general public).

Estates often consist primarily of non-liquid assets. That's normal, because our lives are full of assets of various degrees of liquidity. That's not a problem, unless there's a financial emergency. (Quick! How much cash could you get together within twenty-four hours?)

Death brings urgency to our economic state. Often there's a cash crunch immediately following a death: Assets, assets everywhere, but not enough liquid. There are sudden financial needs—at a time when it can be very difficult to make prudent decisions about finding the cash to meet those needs.

Sure, you can sell assets at price that will guarantee liquidity—if you don't mind losing a lot of value just to make a quick sale. (We're all familiar with liquidation sales: They're quite a bargain if you're buying, but for the people selling the goods the losses can be huge.)

Because of the sudden need for cash, planning for sufficient estate liquidity is critical in many business situations and prudent when trying to prevent an estate from shrinking. After all, if you're not forced to convert assets to cash immediately, you're far more likely to get a good price for them—if you decide to sell.

Sufficient estate liquidity also means that selling will be a choice, not an economic necessity. So your spouse will be able to keep the house and your children will be able to fight over the keys to your car—if you plan for estate liquidity.

Planning for Liquidity

Proper planning should ensure that your estate will have enough cash available to meet all estate planning needs and provide adequate estate liquidity. But what's "adequate"?

To decide what's adequate in your particular situation, take the following steps:

1. Determine potential needs.
2. Assign a realistic value to the needs.
3. Examine the resources that can be used.

Determine Your Potential Needs

First you've got to anticipate the financial demands that your estate is likely to face. An estate may need liquidity to meet any or all of the following needs:

- To pay debts
- To settle creditor claims
- To pay probate and administrative expenses
- To pay funeral and burial expenses
- To cover the income needs of family members
- To pay income and estate taxes
- To ensure smooth transition of business interests
- To execute buy/sell agreements

> **Buy/Sell Agreement:** The most common way to transfer ownership of a business when a partner dies: All partners in a business agree to purchase the interest of any partner who dies.

Calculate the Costs of Your Needs

Which estate needs should you plan to cover? List those needs. How much do you expect each to cost your estate? You may want to make a few phone calls to get those figures.

As you calculate potential costs, you should bear in mind the probable effect of inflation. If you're an optimist, you might want to assume a rate of 3% or 4%. If you're the type who worries a lot, use a higher figure. Then add up all of those projected costs. That total is your liquidity deficiency.

Examine Your Resources

Maybe you've got a pot of cash somewhere that will cover your liquidity deficiency. In that case, congratulations! You can skip on to the next chapter.

If not, then you definitely need to read this next section. To cover your deficiency, we're going to discuss the following common approaches to ensuring estate liquidity:

- Creating a specific savings fund
- Using operating cash on hand
- Selling off assets
- Mortgaging or borrowing
- Using life insurance proceeds

Creating a Specific Savings Fund

Creating a savings fund for estate liquidity requires significant lead time.

Let's assume you establish an account to provide funding for a buy/sell agreement with a target need of $300,000. Your investments allow you to earn an after-tax return of 6% and your cash flow allows you to invest $20,000 annually. How long will it take to fund the buy/sell account? Just a little over eleven years.

(If you want to go through those years one by one and evaluate the progress of your fund, see **Figure 5A**.)

Figure 5A: Analysis of Special Savings Fund

Year	Projected Deposits Annually (a)	Projected Increase at 6% Net (b)	Total Fund Annually (c)	Fund Goal $300,000 Shortage (d)
1	$20,000	$0	$20,000	($280,000)
2	$20,000	$1,200	$41,200	($258,800)
3	$20,000	$2,472	$63,672	($236,328)
4	$20,000	$3,820	$87,492	($212,508)
5	$20,000	$5,250	$112,742	($187,258)
6	$20,000	$6,765	$139,506	($160,494)
7	$20,000	$8,370	$167,877	($132,123)
8	$20,000	$10,073	$197,949	($102,051)
9	$20,000	$11,877	$229,826	($70,174)
10	$20,000	$13,790	$263,616	($36,384)
11	$20,000	$15,817	$299,433	($567)

Key:
(a) $20,000 annually deposited into special savings fund. (b) Assumed rate of return 6% after taxes. (c) Accumulation of fund including growth. (d) Projected shortage of fund versus goal.

Now, what if the need is $1,000,000 and you still allocate $20,000 annually? How long will it take to reach your goal? To make a cool mil, it will take twenty-three years and nine months. (**Figure 5B** shows the annual increments for each one of those years.)

Figure 5B: Analysis of Special Savings Fund

Year	Projected Deposits Annually (a)	Projected Increase at 6% Net (b)	Total Fund Annually (c)	Fund Goal $1,000,000 Shortage (d)
1	$20,000	$0	$20,000	($980,000)
2	$20,000	$1,200	$41,200	($958,800)
3	$20,000	$2,472	$63,672	($936,328)
4	$20,000	$3,820	$87,492	($912,508)
5	$20,000	$5,250	$112,742	($887,258)
6	$20,000	$6,765	$139,506	($860,494)
7	$20,000	$8,370	$167,877	($832,123)
8	$20,000	$10,073	$197,949	($802,051)
9	$20,000	$11,877	$229,826	($770,174)
10	$20,000	$13,790	$263,616	($736,384)
11	$20,000	$15,817	$299,433	($700,567)
12	$20,000	$17,966	$337,399	($662,601)
13	$20,000	$20,244	$377,643	($622,357)
14	$20,000	$22,659	$420,301	($579,699)
15	$20,000	$25,218	$465,519	($534,481)
16	$20,000	$27,931	$513,451	($486,549)
17	$20,000	$30,807	$564,258	($435,742)
18	$20,000	$33,855	$618,113	($381,887)
19	$20,000	$37,087	$675,200	($324,800)
20	$20,000	$40,512	$735,712	($264,288)
21	$20,000	$44,143	$799,855	($200,145)
22	$20,000	$47,991	$867,846	($132,154)
23	$20,000	$52,071	$939,917	($60,083)
24	$20,000	$56,395	$1,016,312	$16,312

Key:

(a) $20,000 annually deposited into special savings fund. (b) Assumed rate of return 6% after taxes. (c) Accumulation of fund including growth. (d) Projected shortage of fund versus goal.

We've got to ask a very delicate question here: How much time do you have? Maybe you've got eleven years. Maybe you've got twenty-three years and nine months. You could have many, many more years than that. But, as we all know, any of us could die at any given moment.

If you evaluate this approach to liquidity, it could make good sense in terms of time and money—in theory. But if you want certainty, never count on time being on your side.

Using Operating Cash on Hand

In most business situations, using operating cash on hand is only a short-term solution. It's usually not long before the appropriation adversely affects the business. You should not count on cash on hand to do more than buy a little time before a permanent solution can be arranged.

In personal situations, cash on hand typically is not adequate to meet estate liquidity needs. Let's consider a variation on that question we asked at the beginning of this chapter: How much cash could your executor get together to settle the estate without losing too much value? Would that amount take care of the needs?

Selling Off Assets

Quite often, the executor of a will may meet the need for liquidity by selling off some assets. The liquidation of assets can be handled in three ways.

If your executor decides to sell personal or business assets that do not have a ready market, he or she will likely be offering those assets to the public at an estate auction or estate sale. That means those assets will probably sell at less than their value, perhaps far less. (What's your first reaction when you're driving or walking down a street and you notice an "Estate Sale" sign? An opportunity to get some bargains, right?)

Your executor may decide to sell at least some assets separately. That should allow him or her to negotiate some better selling. But if the buyers know the reason for the sale, the circumstances may lead them to offer less money than your assets might otherwise have commanded.

Investments may be the best choice for liquidity. Your executor could sell at least certain investments anonymously, so the estate may not lose any of its value. But then the beneficiaries of your will would not enjoy the appreciation on those investments that you were careful to choose and hold.

Obviously this whole area of raising liquidity is a matter that you should discuss with the person you've named to execute your will. Discuss your assets, and then decide what might be best to sell to cover any liquidity deficiency and how the sales might best be handled.

Then crunch the numbers. How much cash could be raised quickly? Would that amount be sufficient to cover the needs of your estate?

Mortgaging or Borrowing

A common solution for estate liquidity problems is to mortgage or borrow. Unfortunately, this can be difficult. In a business situation, it often means borrowing money after the person with the greatest expertise to make money and repay debt is gone. In a personal situation, this is usually a time when taking on more debt is the last thing the survivors want to do.

How much debt might your loved ones or partners incur to provide the necessary liquidity?

Let's look at an example.

If they needed $300,000 and could borrow that amount at 6% interest, over a twenty-year repayment period it would cost them $489,000 (**Figure 5C**).

PLANNING FOR LIQUIDITY

Figure 5C: Analysis of Borrowing $300,000 for 20 Years

Year	Annual Interest Rate 6.00%	Annual Debt Payment	Balance Outstanding Debt	Cumulative Interest Expense (Not Tax-Adjusted)
1	$18,000	$15,000	$285,000	$18,000
2	$17,100	$15,000	$270,000	$35,100
3	$16,200	$15,000	$255,000	$51,300
4	$15,300	$15,000	$240,000	$66,600
5	$14,400	$15,000	$225,000	$81,000
6	$13,500	$15,000	$210,000	$94,500
7	$12,600	$15,000	$195,000	$107,100
8	$11,700	$15,000	$180,000	$118,800
9	$10,800	$15,000	$165,000	$129,600
10	$9,900	$15,000	$150,000	$139,500
11	$9,000	$15,000	$135,000	$148,500
12	$8,100	$15,000	$120,000	$156,600
13	$7,200	$15,000	$105,000	$163,800
14	$6,300	$15,000	$90,000	$170,100
15	$5,400	$15,000	$75,000	$175,500
16	$4,500	$15,000	$60,000	$180,000
17	$3,600	$15,000	$45,000	$183,600
18	$2,700	$15,000	$30,000	$186,300
19	$1,800	$15,000	$15,000	$188,100
20	$900	$15,000		$189,000
Total Payment: $489,000 (Principal and Interest)				

If they needed $1,000,000 and could borrow that amount at 6% interest over a twenty-year repayment period, it would cost them $1,630,000 (**Figure 5D**).

Figure 5D: Analysis of Borrowing $1,000,000 for 20 years

Year	Annual Interest Rate 6.00%	Annual Debt Payment	Balance Outstanding Debt	Cumulative Interest Expense (Not Tax Adjusted)
1	$60,000	$50,000	$950,000	$60,000
2	$57,000	$50,000	$900,000	$117,000
3	$54,000	$50,000	$850,000	$171,000
4	$51,000	$50,000	$800,000	$222,000
5	$48,000	$50,000	$750,000	$270,000
6	$45,000	$50,000	$700,000	$315,000
7	$42,000	$50,000	$650,000	$357,000
8	$39,000	$50,000	$600,000	$396,000
9	$36,000	$50,000	$550,000	$432,000
10	$33,000	$50,000	$500,000	$465,000
11	$30,000	$50,000	$450,000	$495,000
12	$27,000	$50,000	$400,000	$522,000
13	$24,000	$50,000	$350,000	$546,000
14	$21,000	$50,000	$300,000	$567,000
15	$18,000	$50,000	$250,000	$585,000
16	$15,000	$50,000	$200,000	$600,000
17	$12,000	$50,000	$150,000	$612,000
18	$9,000	$50,000	$100,000	$621,000
19	$6,000	$50,000	$50,000	$627,000
20	$3,000	$50,000		$630,000

Total Payment: $1,630,000 (Principal and Interest)

USING LIFE INSURANCE PROCEEDS

Life insurance proceeds are without a doubt the very best way to fund estate liquidity. Why?

For two very good reasons:

- Life insurance provides proceeds when they're needed, regardless of how much time has elapsed.

PLANNING FOR LIQUIDITY

- Life insurance typically costs less to fund than any other method of ensuring liquidity.

Figures 5E and 5F show how a life insurance policy can provide liquidity in our two example amounts, $300,000 and $1,000,000.

Figure 5E: Analysis of Life Insurance Estate Liquidity Requirement $300,000 Funding for Male Age 65, Assumed Policy Rate 6.5%*

Year	Age	Annual Payments to Policy (a)	Cash Value Build-Up (b)	Face Value Of Life Insurance (c)	Annual Loan Payment (d)	Annual Savings Payment (e)
1	65	$13,915	$5,333	$300,000	$24,675	$20,000
2	66	$13,915	$16,686	$300,000	$24,675	$20,000
3	67	$13,915	$28,098	$300,000	$24,675	$20,000
4	68	$13,915	$39,650	$300,000	$24,675	$20,000
5	69	$13,915	$51,498	$300,000	$24,675	$20,000
6	70	$13,915	$64,225	$300,000	$24,675	$20,000
7	71	$13,915	$77,015	$300,000	$24,675	$20,000
8	72	$13,915	$90,272	$300,000	$24,675	$20,000
9	73	$13,915	$103,973	$300,000	$24,675	$20,000
10	74	$13,915	$118,170	$300,000	$24,675	$20,000
11	75	$0	$120,276	$300,000	$24,675	$20,000
12	76	$0	$122,083	$300,000	$24,675	$20,000
13	77	$0	$123,545	$300,000	$24,675	$20,000
14	78	$0	$124,579	$300,000	$24,675	$20,000
15	79	$0	$124,908	$300,000	$24,675	$20,000
16	80	$0	$124,419	$300,000	$24,675	$0
17	81	$0	$123,243	$300,000	$24,675	$0
18	82	$0	$121,240	$300,000	$24,675	$0
19	83	$0	$118,238	$300,000	$24,675	$0
20	84	$0	$114,023	$300,000	$24,675	$0
21	85	$0	$108,991	$300,000	$0	$0
22	86		$102,178	$300,000		
23	87		$93,147	$300,000		
24	88		$81,327	$300,000		
25	89		$65,969	$300,000		
26	90	$0	$46,321	$300,000	$0	$0
27	91		$21,183	$300,000		
28	92		$0	$300,000		

5.11

Year	Age	Annual Payments to Policy (a)	Cash Value Build-Up (b)	Face Value Of Life Insurance (c)	Annual Loan Payment (d)	Annual Savings Payment (e)
29	93		$0	$300,000		
30	94		$0	$300,000		
31	95		$0	$300,000		
32	96		$0	$300,000		
33	97		$0	$300,000		
34	98		$0	$300,000		
35	99		$0	$300,000		
36	100	$0	$0	$0	$0	$0
		Insurance $139,150	Total Costs of Funding for $300,000		Borrowing $493,500	Saving $300,000

Key:
(a) Annual contributions to life insurance contract. (b) Annual projected build-up of cash value in insurance contract. (c) Available cash in the event of death of the insured (death benefit). (d) Annual outlay savings. (e) Annual outlay borrowing.

*Major national insurance company illustration (policy provides 16-year guarantee at illustrated premium).

Figure 5F: Analysis of Life Insurance Estate Liquidity Requirement $1,000,000 Funding for Male Age 65, Assumed Policy Rate 6.5%*

Year	Age	Annual Payments to Policy (a)	Cash Value Build-Up (b)	Face Value of Life Insurance (c)	Annual Loan Payment (d)	Annual Savings Payment (e)
1	65	$43,240	$16,020	$1,000,000	$82,250	$20,000
2	66	$43,240	$52,118	$1,000,000	$82,250	$20,000
3	67	$43,240	$88,442	$1,000,000	$82,250	$20,000
4	68	$43,240	$125,235	$1,000,000	$82,250	$20,000
5	69	$43,240	$162,958	$1,000,000	$82,250	$20,000
6	70	$43,240	$203,412	$1,000,000	$82,250	$20,000
7	71	$43,240	$244,010	$1,000,000	$82,250	$20,000
8	72	$43,240	$286,046	$1,000,000	$82,250	$20,000
9	73	$43,240	$329,447	$1,000,000	$82,250	$20,000
10	74	$43,240	$374,352	$1,000,000	$82,250	$20,000
11	75		$382,052	$1,000,000	$82,250	$20,000

PLANNING FOR LIQUIDITY

Year	Age	Annual Payments to Policy (a)	Cash Value Build-Up (b)	Face Value of Life Insurance (c)	Annual Loan Payment (d)	Annual Savings Payment (e)
12	76		$388,910	$1,000,000	$82,250	$20,000
13	77		$394,798	$1,000,000	$82,250	$20,000
14	78		$399,485	$1,000,000	$82,250	$20,000
15	79		$402,147	$1,000,000	$82,250	$20,000
16	80		$402,477	$1,000,000	$82,250	$20,000
17	81		$400,906	$1,000,000	$82,250	$20,000
18	82		$397,061	$1,000,000	$82,250	$20,000
19	83		$390,486	$1,000,000	$82,250	$20,000
20	84		$380,621	$1,000,000	$82,250	$20,000
21	85		$368,993	$1,000,000		$20,000
22	86		$352,662	$1,000,000		$20,000
23	87		$330,536	$1,000,000		$20,000
24	88		$301,213	$1,000,000		$20,000
25	89		$262,874	$1,000,000		$20,000
26	90		$213,819	$1,000,000		$20,000
27	91		$151,264	$1,000,000		$20,000
28	92		$73,313	$1,000,000		$20,000
29	93		$0	$1,000,000		$20,000
30	94		$0	$1,000,000		$20,000
31	95		$0	$1,000,000		$20,000
32	96		$0	$1,000,000		$20,000
33	97		$0	$1,000,000		$20,000
34	98		$0	$1,000,000		$20,000
35	99		$0	$1,000,000		$20,000
36	100		$0			$20,000
		Insurance $432,400	Total Costs of Funding for $1,000,000		Borrowing $1,645,000	Saving $720,000

Key:
(a) Annual contributions to life insurance contract. (b) Annual projected build-up of cash value in insurance contract. (c) Available cash in the event of death of the insured (death benefit). (d) Annual outlay savings. (e) Annual outlay borrowing.

*Major national insurance company illustration (policy provides 15-year guarantee at illustrated premium).

There may be a disadvantage, however, to depending on life insurance. Taking out a policy requires moderately good health. In the past that may have put this strategy out of reach for some people.

Fortunately, a relatively recent creation has improved this situation for married people—*survivorship life insurance*. Also known as joint, joint survivorship, two-life, or second-to-die, this type of policy was first developed in 1961, but gained popularity with the passing of the Economic Recovery Tax Act of 1981.

Survivorship life insurance is a single policy that insures two lives, usually spouses, and pays off only upon the death of the second person. In some instances, only one person must be insurable for both to qualify for a survivorship policy. This type of life insurance may be especially useful in estate planning for families where one person has serious health problems.

> **Survivorship Life Insurance:** A life insurance policy that covers two people, usually spouses, and pays off only when the second person dies. Also known as joint life, two-life, or second-to-die.

Survivorship insurance not only allows coverage for people who might have trouble getting conventional life insurance, but it also can provide coverage for less expense.

Premium rates are based on the joint life expectancy of the two persons, so for a given face value the premium rate is lower than it would be for either person individually. A survivorship policy for $1,000,000 costs less than a conventional $1,000,000 policy on one person or two separate $500,000 policies. Survivorship life insurance is available as whole or universal life.

Figures 5G and 5H show how a survivorship life insurance policy can provide liquidity in our two example amounts, $300,000 and $1,000,000.

PLANNING FOR LIQUIDITY

Figure 5G: Analysis of Life Insurance Estate Liquidity Requirement $300,000 Funding for Two Lives Male and Female Age 65, Assumed Policy Rate 6.85%*

Year	Age	Annual Payments to Policy (a)	Cash Value Build-Up (b)	Face Value of Life Insurance(c)	Annual Loan Payment (d)	Annual Savings Payment (e)
1	65	$10,000	$7,626	$300,000	$24,675	$20,000
2	66	$10,000	na	$300,000	$24,675	$20,000
3	67	$10,000	na	$300,000	$24,675	$20,000
4	68	$10,000	na	$300,000	$24,675	$20,000
5	69	$10,000	$42,261	$300,000	$24,675	$20,000
6	70	$10,000	na	$300,000	$24,675	$20,000
7	71	$10,000	na	$300,000	$24,675	$20,000
8	72	$10,000	na	$300,000	$24,675	$20,000
9	73	$10,000	na	$300,000	$24,675	$20,000
10	74	$10,000	$108,304	$300,000	$24,675	$20,000
11	75	$0	na	$300,000	$24,675	$20,000
12	76	$0	na	$300,000	$24,675	$0
13	77	$0	na	$300,000	$24,675	$0
14	78	$0	na	$300,000	$24,675	$0
15	79	$0	$135,423	$300,000	$24,675	$0
16	80	$0	na	$300,000	$24,675	$0
17	81	$0	na	$300,000	$24,675	$0
18	82	$0	na	$300,000	$24,675	$0
19	83	$0	na	$300,000	$24,675	$0
20	84	$0	$171,945	$300,000	$24,675	$0
21	85	$0	na	$300,000	$0	$0
26	90	$0	$226,654	$300,000	$0	$0
31	95	$0	$316,411	$316,411	$0	$0
36	100	$0	$454,302	$454,302	$0	$0
	Insurance $100,000		Total Costs of Funding for $300,000		Borrowing $657,280	Saving $220,000

Key:
(a) Annual contributions to life insurance contract. (b) Annual projected build-up of cash value in insurance contract. (c) Available cash in the event of death of the insured (death benefit). (d) Annual outlay savings. (e) Annual outlay borrowing.
*Major national insurance company illustration (policy provides 19-year guarantee at illustrated premium).

5.15

Figure 5H: Analysis of Life Insurance Estate Liquidity Requirement $1,000,000 Funding for Two Lives Male and Female Age 65, Assumed Policy Rate 6.85%*

Year	Age	Annual Payments to Policy (a)	Cash Value Build-Up (b)	Face Value of Life Insurance (c)	Annual Loan Payment (d)	Annual Savings Payment (e)
1	65	$25,484	$9,141	$2,230,000	$82,250	$20,000
2	66	$25,484	$31,946	$2,230,000	$82,250	$20,000
3	67	$25,484	$56,015	$2,230,000	$82,250	$20,000
4	68	$25,484	$81,407	$2,230,000	$82,250	$20,000
5	69	$25,484	$108,186	$1,000,000	$82,250	$20,000
6	70	$25,484	$136,412	$1,000,000	$82,250	$20,000
7	71	$25,484	$166,136	$1,000,000	$82,250	$20,000
8	72	$25,484	$197,426	$1,000,000	$82,250	$20,000
9	73	$25,484	$230,336	$1,000,000	$82,250	$20,000
10	74	$25,484	$264,924	$1,000,000	$82,250	$20,000
11	75		$279,185	$1,000,000	$82,250	$20,000
12	76		$293,843	$1,000,000	$82,250	$20,000
13	77		$308,786	$1,000,000	$82,250	$20,000
14	78		$323,837	$1,000,000	$82,250	$20,000
15	79		$338,760	$1,000,000	$82,250	$20,000
16	80		$353,282	$1,000,000	$82,250	$20,000
17	81		$367,344	$1,000,000	$82,250	$20,000
18	82		$380,767	$1,000,000	$82,250	$20,000
19	83		$393,395	$1,000,000	$82,250	$20,000
20	84		$404,956	$1,000,000	$82,250	$20,000
21	85		$415,120	$1,000,000		$20,000
22	86		$423,564	$1,000,000		$20,000
23	87		$429,870	$1,000,000		$20,000
24	88		$433,533	$1,000,000		$20,000
25	89		$433,961	$1,000,000		$20,000
26	90		$430,828	$1,000,000		$20,000
27	91		$423,282	$1,000,000		$20,000
28	92		$410,393	$1,000,000		$20,000
29	93		$390,985	$1,000,000		$20,000
30	94		$363,562	$1,000,000		$20,000
31	95		$326,170	$1,000,000		$20,000
32	96		$275,101	$1,000,000		$20,000

PLANNING FOR LIQUIDITY

Year	Age	Annual Payments to Policy (a)	Cash Value Build-Up (b)	Face Value of Life Insurance (c)	Annual Loan Payment (d)	Annual Savings Payment (e)
33	97		$206,360	$1,000,000		$20,000
34	98		$113,774	$1,000,000		$20,000
35	99		$1,036	$1,000,000		$20,000
36	100	$0	##	##	$0	$20,000
		Insurance $254,840	Total Costs of Funding For $1,000,000		Borrowing $2,190,940	Saving $720,000
			Key:			

(a) Annual contributions to life insurance contract. (b) Annual protected build-up of cash value in insurance contract. (c) Available cash in the event of death of the insured (death benefit). (d) Annual outlay savings. (e) Annual outlay borrowing.

*Major national insurance company illustration (policy provides 18-year guarantee at illustrated premium).

So, the best way to ensure estate liquidity is through life insurance—conventional or survivorship. However, life insurance should not have your estate named as the beneficiary. This action will guarantee probate and tie up the proceeds when they are needed. Life insurance does not have to be a probate asset. By naming your business partner, your spouse, your loved ones, etc., as beneficiary of your life insurance contract, probate is avoided and the proceeds are available immediately for the intended purpose.

The bottom line is that with careful planning to reduce estate taxes and probate expenses, and a life insurance policy to cover your estate liquidity needs, you can pass your estate intact to your heirs.

LENDER-FUNDED PREMIUM FINANCING

Because it enables a person to obtain an appropriate life insurance policy without depleting cash reserves or liquidating high-performing investments, premium financing can help to provide financial security while meeting your estate-planning goals.

How Does Premium Financing Work?

Upon securing loan approval from the premium finance company, the life insurance policy is most commonly issued into an irrevocable life insurance trust, or ILIT. The borrower will name beneficiaries of the trust, which may include a spouse, children, other natural heirs, business partners, or a charity.

Usually, the borrower will have to post collateral for the loan in the form of the policy itself (through a collateral assignment) and possibly additional collateral for a percentage of the loan amount (on average 25%) in the form of:

- A letter of credit
- Personal guarantee
- Marketable securities
- Cash

Upon securing the required collateral, the initial loan amount is dispersed to the ILIT, which in turn pays the first premium on the life insurance policy. As future premium payments to the insurance company are required, the premium finance company continues to disperse funds according to the premium schedule.

The grantor makes annual gifts of the loan interest payments to the trust. The trust pays the interest to the lender. Interest is NOT deductible by the trust.

Who Qualifies for Premium Financing?

Although qualifying parameters vary between premium financing companies, generally an individual over the age of 35 seeking a life insurance policy with a death benefit of $5 million or more and an annual premium over $100,000 will qualify.

The loan arrangement may last from one year to the life of the policy. The lender provides a loan to the trust annually. Then the premium finance company pays the insurance premium and bills the individual or company, usually in monthly installments, for the cost of the loan.

Benefits of Premium Financing

By borrowing the premiums, qualified individuals will pay as little out of pocket as possible for life insurance. Estate taxes are also affected because instead of gifting large premiums to the trust, the premium dollars are being loaned to the trust.

Additionally, financing premiums:

- Eliminates the requirement for a large up-front payment to an insurance company.

- Enables multiple insurance policies to be attached to a single premium finance contract, allowing for a single payment plan to cover all insurance coverage.

- Offers transparency to the individual or company insured. Brokers transmit completed premium finance agreements to the premium finance company, and policyholders are billed as they would be for any other typical insurance policy.

- Allows clients to obtain needed coverage without liquidating other assets.

Types of Premium Financing

- **Traditional:** The client enters a fully collateralized loan arrangement with the intention of holding the life insurance policy to maturity. Traditional financing arrangements are generally purchased for estate liquidity needs and offer the most advantageous loan rates, fees, and spreads. The client may have an exit strategy using other assets in the estate. Traditional finance is particularly effective for clients who have a large but illiquid net worth.

- **Loan Repayment:** When the loan matures, the policy owner may have a number of choices including:
  - Continuing to finance the premiums
  - Repaying the loan plus interest
  - Keeping the life insurance policy

Premium loans can be repaid:
- During life, from the borrower's available assets
- During life, from policy cash values, or
- At death, from policy proceeds

When the loan period ends or the insured passes away, the principal amount plus interest is paid back to the premium finance company and the owner retains the life insurance policy or collects the death benefit to be paid to the policy beneficiary less the outstanding loan value.

If the insured passes away during the financing period, the ILIT will claim the death benefit and disperse the proceeds less the outstanding loan value to the beneficiary. The trust distributes and manages the net death proceeds as specified in the trust document.

Risks of Premium Financing

While the leverage can be a great tool, benefits must be balanced against the following significant risks:

- **Policy Design Risk:** The cost of an increasing death benefit can have a considerable effect on the policy's premium requirements, particularly in the later years. Higher premiums mean larger loans since a larger loan is needed to pay the policy's higher premium. However, a larger loan means that even higher death benefits are needed, which in turn means higher premiums. This circular dependency can require significantly more insurance to satisfy the ultimate loan.

- **Interest Rate Volatility Risk:** Since the interest rate on the premium financing loan is tied to an index, usually the LIBOR (London Interbank Offered Rate), if interest rates rise, the total interest charge will rise as well. If the loan interest rates increase more than projected, clients could be required to pay more money into the program and/or provide more collateral than originally anticipated. If they don't have sufficient funds and collateral to make up this shortfall, the entire loan could be called—which would force them to repay the loan before they planned to.

- **Renewal Risk:** Loans can be made for a fixed term of years, but cannot be made in perpetuity. The lender has the right to call the loan at the end of the term. Virtually all premium financing loans have terms of less than the life of the policy. Premium financing programs assume the loan continuously gets renewed at the end of each term until the client's death or when the client can pay the balance through a liquidation event. Since each loan renewal is subject to 1) the lender's underwriting guidelines, 2) the lender's desire to continue funding insurance premiums, and 3) the client's financial situation, there is no guarantee the lender will renew the client's loan or that any lender will offer a new loan to continue the program.

- **Carrier Credit Rating Risk:** Financing terms are sensitive to the credit rating of the carrier holding the financed policy. Carrier downgrades may result in the lender choosing to not pay additional premiums and/or to call the collateral for the loan.

- **Crediting Rate Risk:** Carriers choose the crediting rate of in-force blocks of business at will. Current crediting rates are not guaranteed. As such, any advantage in interest rates between the policy crediting rate and the loan interest rate may not exist in the future.

- **Collateral Risk:** Collateral requirements may vary with economic conditions and could force the client to liquidate possessions to post collateral. If the value of the client's collateral (such as real estate or securities) falls below the level required by the lender to satisfy the loan, the client could be called upon to provide additional collateral.

- **Gift Tax Risk:** If the lender exercises its rights to take a loan or partial surrender that causes the policy to be at risk of terminating, the grantor would have to transfer additional assets to the trust to prevent lapse. Such transfers could be significant and would be considered gifts to the trust and subject to gift tax.

- **Earnings Risk:** If the policy's death benefit does not, or cannot, grow sufficiently to keep pace with the outstanding loan, then the client is at risk of either not getting as much coverage as expected after the loan is paid off, or, worse, getting no insurance coverage at all and having to come up with additional funds to repay the balance of the loan.

- **Settlement Risk:** Some premium financing programs are sold under the assumption that the policy will have a substantial market value at the end of the term. The client can then exit the financing arrangement and realize a gain on investment. The secondary life insurance market is highly volatile. Settlement offers will vary with the interest rate environment and the degree to which capital will "wait" for a return.

POST ECONOMIC-CRISIS ENVIRONMENT

Premium financing arrangements are currently under intense pressure. Lenders often obtain capital from hedge funds and private sources, which are particularly sensitive to volatility in the credit markets. Many lenders have ceased financing or added substantial fees to new programs. In-force financed policies are being called for collateral in large numbers. Clients who are "underwater," where the loan balance exceeds the policy cash value, are being forced to post additional collateral at low risk–weighted rates and/or surrender the policies and pay the outstanding loan balance out of pocket. Additionally, several carriers who were active in the financing marketplace have been downgraded, causing large-scale exchanges or surrenders from in-force policies.

6

How to Use Trusts in Estate Planning

Many people understand the concept of a will but not the concept of a trust. A major distinction is that a will manages assets only after death. A trust, on the other hand, can provide for the management of assets during life.

Modern estate planning and trusts go hand in hand. Wills usually distribute only part of our estate assets. We can create trusts to receive, manage, and distribute all that we own. In this chapter we will explore both basic trusts and several specific trusts that are used in estate planning.

> **Trust:** A legal entity that can own assets.

What Is A Trust?

In simple terms, a trust is an entity created to own assets. The assets can be of various types—cash, stocks, bonds, real estate, business interests, collections, and almost any other sort of tangible assets.

There are typically three parties to the trust:

Grantor: The person who sets up the trust, names the beneficiary and the trustee, and transfers the assets to the trust.

Beneficiary: The person or institution named to receive the benefits of the trust.

Trustee: The person or institution named to manage the trust.

Figure 6A shows the basic dynamics of a trust.

Figure 6A: Basic Trust Dynamics

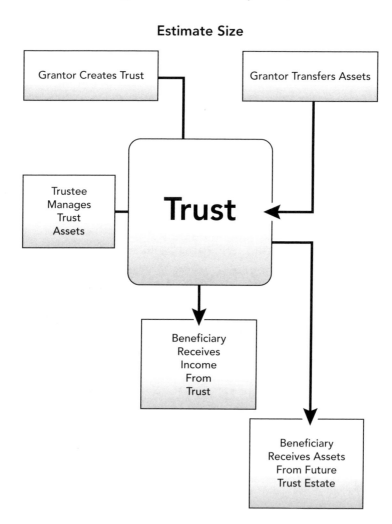

Let's now define trusts more completely. A trust is a formal written agreement by which a person (the grantor) enables a person or an institution (the trustee) to hold and manage assets for the benefit of a person or an institution (the beneficiary) in accordance with the instructions in the trust agreement.

In explaining the dynamics of a trust, we've kept the three parties in the singular—grantor, trustee, beneficiary. A grantor can name more than one trustee and/or more than one beneficiary.

In setting up the trust, the grantor provides written instructions for the trustee concerning how he or she is to manage or distribute the trust assets. These instructions can be quite basic or very specific.

Trusts can be fairly simple or very intricate, depending on the purposes and the circumstances. But the basic trust dynamics remain constant.

Types of Trusts

There are two main types of trusts—*living* trusts and *testamentary* trusts.

A *living* trust is any trust that you create during your lifetime. A living trust is also called the inter vivos trust (a Latin term meaning "between or among the living"). This type of trust can be funded (assets transferred to the trust) or not funded (left empty until needed), depending on what the trust is meant to do.

A *testamentary* trust is created by your will. It really isn't a trust at all until after death. Then the provisions that are written into the will allow for the type of trust to be set up and specify how it is to be funded.

> **Living Trust:** Any trust that you create during your lifetime. Also called an inter vivos trust.
>
> **Testamentary Trust:** Any trust that you create through your will.

The distinction between living trusts and testamentary trusts is basically a matter of life and death. That seems like a very important difference, doesn't it? But many types of trusts can be created as either living trusts or testamentary trusts—with some differences. So, here at least, the difference between life and death can be of little importance.

Far more important in practical terms are the distinctions among the various types of trusts in terms of what they do and how they work. In this chapter we'll discuss the types of trusts that will be most useful to you in estate planning.

Basic Living Trust

Think about the three players in the trust game—the grantor, the beneficiary, and the trustee. In a basic living trust, the grantor can play all three roles. (That would be impossible, of course, in a testamentary trust!)

A person can set up a trust that he or she manages as trustee and/or from which he or she receives all the benefits as beneficiary. Since nobody lives forever, the grantor also names a second trustee and a second beneficiary. This trustee and this beneficiary are called *contingent* because they assume those roles only when the grantor dies.

> **Contingent Beneficiary:** Person named to succeed the grantor of a living trust who has named himself or herself as beneficiary.
>
> **Contingent Trustee:** Person named to succeed the grantor of a living trust who has named himself or herself as trustee.

Upon the death of the grantor, the contingent trustee steps in to manage or distribute assets and the contingent beneficiary begins to receive the assets from the trust. In a simple living trust, that may happen immediately.

However, the grantor may have left trust instructions for the trustee to hold assets for a period of time before distributing any or all of them. This would be the case, for example, if the grantor named minor children as contingent beneficiaries of the trust.

The transfer of legal roles from the grantor as trustee and beneficiary to the contingent trustee and the contingent beneficiary happens without probate involvement. In fact, many living trusts are set up for the sole purpose of keeping estate assets from going through probate.

But death is not the only contingency for which you can plan by naming a contingent trustee. What if you become disabled or incapacitated? Who would manage your assets?

The living trust allows you to plan for such an unfortunate event. If you place those assets in a living trust and provide for the possibility of disability or incapacitation, your contingent trustee can take over the management of the trust assets without any involvement of the court system.

If you want to control how your assets are handled while you're alive or after your death, you can do so with a living trust. The flexibility of the living trust is restricted only by your imagination and goals. There are some legal restrictions, of course. You should use a qualified attorney to draft the trust document.

A word of caution about living trusts: A trust can manage only assets that it owns! A very common mistake is to have an attorney draft a living trust and then not title assets to the trust or to purchase new assets and forget to tide them to the trust. If you want a trust to manage assets, the trust must own those assets.

Pour-Over Will

A document that typically accompanies a living trust is the *pour-over will*. This is a will that directs property to go to another legal entity, usually a trust. The process is simple: You set up a living trust and then include a provision in your will that names your trust as beneficiary of the residuary estate—something like "I bequeath my residuary estate to The J.J. Jones Trust." Then, upon your death, the executor of your will takes care of your specific bequests and pays death taxes and claims against the estate, then pours over the rest of the estate into the trust.

Pour-Over Will: A will or a provision in a will that directs property to go to another legal entity, usually a trust.

Pour-over wills do not avoid probate. They simply direct the distribution of some probate estate assets into a living trust.

KEEPING THE FARM IN YOUR FAMILY

CREDIT SHELTER TRUST

The credit shelter trust provides a way to use the unified credits of both husband and wife and to generate income for the surviving spouse. (This trust is known under various names—credit shelter family trust, family trust, bypass trust, B trust, family credit shelter trust, credit trust, and exemption trust.)

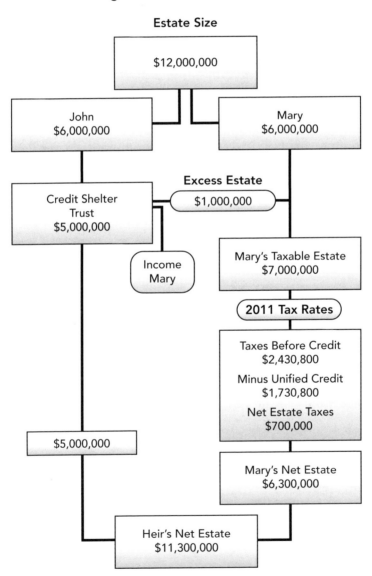

Figure 6B: Credit Shelter Trust

6.6

The credit shelter trust is perhaps the most common type of trust used in estate planning. It has numerous variations.

It is customarily designed to receive the maximum amount that can pass free of federal estate tax upon the death of the first of two spouses to die. The grantor has complete control over where the trust assets are ultimately distributed. The terms of the trust frequently provide for all of its income to be payable to the surviving spouse for life. It can also distribute principal if the terms so indicate and certain criteria are met. Then, when the second spouse dies, the assets pass directly or through a trust for one or more beneficiaries, according to the instructions of the first spouse, bypassing the estate tax.

Credit Shelter Trust is a trust that reduces estate taxes by using the unified credits of both husband and wife and generates income for the surviving spouse. Also known as credit shelter family trust, family trust, bypass trust, B trust, family credit shelter trust, credit trust, and exemption trust.

Note: A grantor can fund a credit shelter trust only with assets that he or she owns separately. So if you and your spouse own everything jointly, you'll need to retitle the assets that you want to place in the trust.

Qualified Terminal Interest Property (QTIP) Trust

A common type of trust used in estate planning is the qualified terminal (or terminable) interest property (QTIP) trust. The QTIP trust is often a testamentary trust, set up for the benefit of a spouse.

An effective way for a married couple to minimize estate tax is by using the unified credit exemption for both husband and wife. However, one of the problems in that example is that Mary didn't receive any income from John's estate.

The QTIP trust takes advantage of the unlimited marital deduction, so when it receives assets, it defers potential estate taxes. It also provides income for the surviving spouse. A QTIP trust must meet two criteria:

1. It must distribute trust income to the surviving spouse during his or her lifetime. (It can also distribute principal if the terms of the trust so indicate and if certain criteria are met.)

2. It must be subject to estate taxes when the surviving spouse dies. The assets are then distributed according to the instructions of the first spouse.

None of the principal of the trust may pass to anyone other than the spouse during his or her lifetime. At the second spouse's death, the assets in the QTIP trust are included in calculating the estate tax on the second spouse's estate. What remains after paying any estate taxes is distributed in accordance with the provisions of the trust as established by the grantor.

Qualified Terminal (or terminable) Interest Property (QTIP) Trust: A trust that takes advantage of the marital deduction, provides income for the surviving spouse for life, and allows the grantor to determine to whom the trust assets will pass when the surviving spouse dies.

When creating hypothetical examples, this is very easy to do. In the real world, the types of assets that make up an estate are unique and how they are titled will have a great impact on how well an estate plan will work. Balancing estate assets between husband and wife can be very challenging. Meticulously maintaining the proper balance of estates and titling allowing for proper distribution will determine the effectiveness of your estate plan.

In our example shown in **Figure 6C**, John's and Mary's estates use both of their unified credit/estate exemptions **($345,800)**. Mary receives the income from both trusts and there is no estate tax erosion of assets until she dies. The impact of the estate tax under this combination is essentially the same as **Figure 4B**, but John determines the distribution of his entire estate by setting up the QTIP trusts and the credit shelter trust (see **Figure 6C**).

Figure 6C: Credit Shelter Family Trust / QTIP Trust

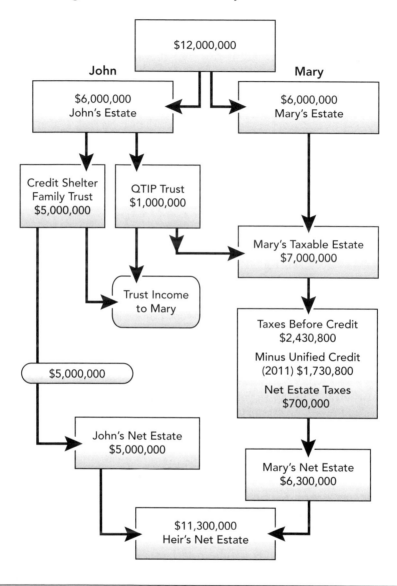

Qualified Domestic Trust

Most people don't know about the qualified domestic trust. In fact, there's no reason for many people to care about it. In fact, if you and your spouse are both citizens of the United States, skip ahead to the next section.

The Qualified Domestic Trust (QDOT) is a special form of QTIP trust for a spouse who is not a U.S. citizen. Under federal estate tax law, transfers of property to a spouse who is not a U.S. citizen do not qualify for the marital deduction unless they are in the form of a trust.

The QDOT keeps any spouse who is not a citizen from receiving assets from a citizen spouse free of estate taxes, and then taking the assets back to his or her home country where they will never be subject to U.S. estate tax. Uncle Sam simply wants to make sure that he collects estate taxes from everybody.

A Comment About Life Insurance

There are many good reasons for life insurance, whatever your age. Life insurance can provide immediate cash or an income stream for your family. Life insurance can also provide liquidity to pay estate taxes to fund a business buy/sell arrangement.

Yet there's a point that many people misunderstand about estate planning: Although a life insurance death benefit is not subject to income tax, it's included in your taxable estate if you own the policy. This point is so often missed that it can hardly be overstated.

Let's take the situation of John and Mary as an example. Assume that John owns a $500,000 life insurance policy on himself. This death benefit would increase the size of his estate to $1,500,000. Let's take a look at the impact that the benefit would have on their combination of credit shelter trust and QTIP trust in their estate planning (see **Figure 6D**).

The impact that the benefit would have on a combination of credit shelter trust and QTIP trust in their estate planning is that the estate tax jumped by $550,000 or 55% of the $1,000,000 death benefit! In other words, Uncle Sam collected almost half of the life insurance proceeds—even though he was not named on the policy as a beneficiary! The eventual net death benefit from the insurance policy for the heir was only $450,000. Ouch!

HOW TO USE TRUSTS IN ESTATE PLANNING

Figure 6D: Impact on Estate Planning

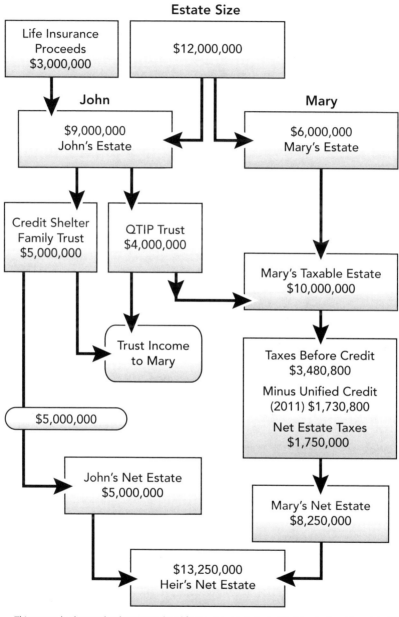

This example shows what happens when life insurance is inducted inside of a taxable estate. The policy for $3,000,000 is owned by John at his death. John and Mary's estate is subject to estate taxes. By owning the life insurance, the proceeds of the death benefit are included in John's estate for estate tax purposes. The incremental tax attributable to the life insurance is $1,050,000.

That's because, if you own a life insurance policy, the proceeds are included in your estate when determining estate taxes, although they're not subject to probate proceedings unless you name your estate as the beneficiary. Although a life insurance policy can be an excellent idea, the proceeds can substantially increase the value of your estate and the amount of estate taxes that must be paid at some point.

Irrevocable Life Insurance Trust (ILIT)

There's a way to prevent Uncle Sam from becoming a beneficiary of your life insurance policy. You can set up an irrevocable life insurance trust (ILIT) to own the policy and keep the death benefit proceeds outside your taxable estate. This trust can save a family tens of thousands of dollars in estate taxes.

The method for paying premiums from the ILIT trust is to make gifts to the trust for beneficiaries.

Special features of the trust cause these gifts to be "present interest" gifts, which allow the transfers to the trust to be exempt under the $11,000 annual gift rule.

> **Irrevocable Life Insurance Trust (ILIT):** A trust that owns a life insurance policy, so that death benefit proceeds do not enter the estate and get taxed.
>
> **Present Interest Gift:** A gift that the recipient has the right to use immediately.

Figure 6E shows how John and Mary can set up an ILIT and name their heir(s) as beneficiary, so that the life insurance proceeds pass directly to the heirs. That way, their heirs benefit from the entire $500,000 in death proceeds without any increase in estate tax liability.

You can easily understand from our example how an ILIT, by protecting life insurance proceeds against estate taxes, can be a powerful tool in estate planning. A life insurance policy owned by an ILIT can provide instant liquidity and eliminate estate shrinkage caused by estate taxes and other final expenses.

HOW TO USE TRUSTS IN ESTATE PLANNING

Figure 6E: ILIT Trust (2011 Tax Rates)

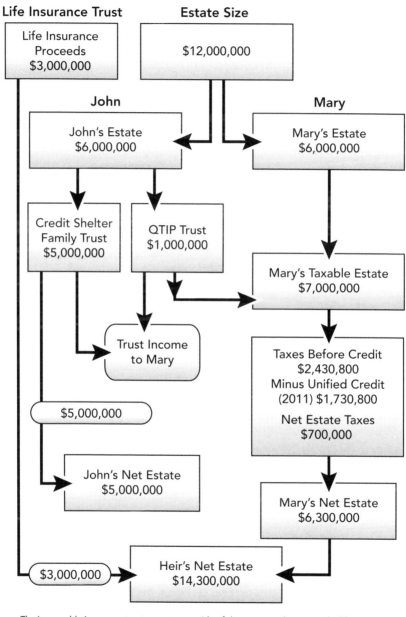

The irrevocable insurance trust operates outside of the estate and protects the life insurance proceeds $3,000,000 from being estate taxed. The proceeds are 100% available for estate liquidity and final expenses.

6.13

We offer you three pieces of advice here.

First, don't wait to title your life insurance policy in an ILIT or you might get caught by the three-year rule. What's that? Well, if a person dies and has owned what the legal folks call "incidents of ownership" in a life insurance policy during the last three years of his or her life, the proceeds are included in that person's gross estate. This rule is intended to prevent terminally ill individuals from avoiding estate tax on insurance benefits by transferring ownership of a policy immediately prior to death. So, if you want to take advantage of an ILIT, don't delay.

Second, keep in mind that transferring ownership of life insurance to an irrevocable trust may result in a gift tax. The gift value is the total of premiums you've paid so far or the current cash value of the policy, depending on the type of policy. If the value is $11,000 or less, this transfer can come in under the annual limit for gifts.

> **Incidents of Ownership:** All or any control over a life insurance policy, including the rights to borrow against the policy or change the beneficiary.

Third, since the ILIT is an irrevocable trust, you give up all of your ownership rights. That means that you cannot borrow against your policy or change your beneficiaries.

Now you know the advantages and the disadvantages of an ILIT. You may want to consult an estate planner. For a single person with an estate of more than $1,000,000 (2011) including life insurance or a married couple with an estate over $2,000,000 (2011), an irrevocable life insurance trust may be appropriate.

Generation-Skipping Transfer Tax

Historically, wealthy families have used trusts and other techniques to pass property down through several generations without paying estate tax. How does this work?

Here's an example. A man creates a trust and names his son and his grandchildren as beneficiaries—his son to receive income from the trust during his life and then his grandchildren to receive the assets held in the trust. The advantage of this planning would be that it provides some financial benefit from the trust assets to one generation without incurring any estate tax upon the transfer to the next generation. Because the son in our example has no right to the trust assets, when he dies the assets are not included in his taxable estate.

This is what's known as a generation-skipping transfer, because the assets in the trust go from the grantor to his grandchildren, skipping his son. The trust that allows this transfer is called, as you might expect, a generation-skipping trust.

We began this section with an important word—"historically." That's because generation-skipping trusts are largely a thing of the past.

Why use a generation-skipping trust? To avoid estate taxes. So, how does Uncle Sam feel when people find legal ways to avoid paying taxes? He creates more laws.

> **Generation-Skipping Transfer:** The passing of assets from the owner to his or her grandchildren, so that they are never in the possession of the owner's child or children.
>
> **Generation-Skipping Trust:** A trust that allows assets to bypass a generation, so that grandchildren receive property directly from their grandparents, without it passing through their parents.

The law created in this case (passed in 1976, then repealed, and finally reenacted in 1986) imposes a separate estate tax structure known as the generation-skipping transfer (GST) tax. The GST tax is intended to complement the federal estate tax by restricting taxpayers from transferring wealth to successive generations free of estate taxes. So, very wealthy families can no longer use generation-skipping trusts to keep very big estates from a very big tax bite.

But there's still a little relief possible through generation-skipping trusts, as we'll explain shortly.

The GST Tax is applied to:
- Direct skips
- Taxable distributions
- Taxable terminations

OK. What does that mean in English? If you make a transfer directly to a descendant who is more than one generation below you (your grandchildren or great-grandchildren or beyond), the GST Tax will apply. That's known as a *direct skip*.

If you create a trust for a child and name grandchildren as secondary beneficiaries, any distributions made from the trust to the grandchildren will be subject to the GST Tax, one way or the other. If the trust distributions to the grandchildren come while their parent is alive, they're called *taxable distributions*. If the trust distributions to the grandchildren come when the trust terminates, they're called *taxable terminations*.

Now you understand how the generation-skipping transfer tax works. So, what's the bottom line?

The GST Tax is substantial—the tax rate is up to 55%, the highest estate and gift tax rate. Ouch!

But the generation-skipping trust is not completely a thing of the past. It remains alive, to some extent at least, through the generation-skipping tax exemption.

Every individual is allowed a GST Tax exemption of $1,000,000. This amount will be indexed for inflation. This exemption may be allocated to property or assets that a person transfers while alive or when he or she dies. A married couple may combine the two exemptions and transfer up to $2,000,000 (2011) without paying the GST Tax.

The very wealthy can no longer avoid hefty estate taxes by using generation-skipping trusts. However if you have a large estate, you should certainly consider using your GST Tax exemption.

A Trust That Is Not a Trust

Since the focus of this chapter is trusts, we should include here a few words about something that's called a trust but is not technically a trust—the Totten trust.

A Totten trust is actually a bank account that the owner titles in such a way that, upon his or her death, the contents pass to one or more other named beneficiaries without going through probate.

Once you sort through all the various names, this "trust" is actually very simple to understand and use.

You open a bank account in your name, as depositor and as "trustee for the benefit of beneficiary." The person whom you designate as beneficiary will assume ownership of the account when you die. However, as long as you live, the beneficiary has no rights to the account. You can do with it as you like—even spend all the money in the account. Upon your death, the beneficiary simply presents the bank with proof of his or her identity and a certified copy of your death certificate.

> **Totten Trust:** A shared bank account that belongs to the depositor until he or she dies, then passes to the designated beneficiary. Also known as a payable-on-death (or pay-on-death) or POD account, informal trust, or bank trust account.

If you'd like to designate more than one beneficiary, the process is the same, except that after you name the beneficiaries, you add the wording "equal beneficiaries." That way, your beneficiaries will each enjoy an equal share of the account.

It's as simple as that. Well, in principle. If the beneficiary dies before the depositor and there's no successor beneficiary named, the money in the account transfers as part of the residuary estate, according to the terms of the will for the distribution of the residuary estate.

The benefit of a Totten trust is that the money in the account bypasses probate. However, the account is included in the taxable estate of the depositor.

Although this "trust" is usually set up as a new account, you can also add a trust designation to a current account. Totten trust accounts are usually savings accounts or certificates of deposit accounts, but they can be checking accounts as well. It's easy to open a Totten trust. Most banks and credit unions have standard forms and can handle the paperwork for you.

7

Planning: "Just In Case…"

As you have discovered, just having a will does not mean you have a complete estate plan. A will is activated only when you die. What happens if you become incapacitated and are unable to make decisions and manage your affairs?

It's a disturbing thought. Nobody feels comfortable thinking about living with diminished abilities. But that's what happens to thousands of people. Planning now can substantially eliminate problems, both for you and for your loved ones.

Good estate planning involves more than passing on your assets to those you leave behind when you die. It should also involve taking actions while you are in good physical and mental health so that somebody can act on your behalf—to make financial, legal, and medical decisions—if you become incapacitated and are no longer able to make decisions.

> **Incapacitation:** In general, the loss of mental competence, the inability to make decisions.

A good estate plan covers the possibility of incapacitation. Otherwise, it may become unnecessarily difficult on the members of your family to pay bills, manage financial affairs, make investment decisions, and possibly run a business.

If you don't make the appropriate arrangements in advance, your loved ones may be forced to petition the court to appoint a guardian or a conservator. That's a tough process in several ways.

Conservatorship and Guardianship

If a person becomes incapacitated and incapable of making personal or financial decisions, a court may appoint a conservator or a guardian to manage his or her personal affairs, estate, or both.

> **Guardian:** A person who's legally responsible for managing the affairs and the care of a minor or a person who's incompetent; in some states a conservator plays a similar but more limited role.
>
> **Conservator:** A person who's legally responsible for managing the financial affairs of a person who's incompetent, playing a role similar to that of a guardian.

Guardianship and conservatorship are closely related concepts that vary according to the laws of each state. Here are the basics:

A guardian is a person or an entity appointed to exercise a certain list of powers, including paying for support, maintenance, and education, paying lawful debts, possessing and managing the estate, collecting debts, and instituting lawsuits on behalf of another person, the ward.

A conservator, also known as a limited guardian, is a person or an entity appointed by the court to exercise some, but not all, of these powers on behalf of another person, known as the conservatee.

To create a guardianship or a conservatorship, any person can petition a probate court.

Usually this person is a family member or close friend who's concerned about the person's competence to manage property or make personal decisions.

The petition must set forth the reasons why a guardianship or conservatorship is needed. The petitioner has the burden of proving the person's incapacity. In making its decision, the court applies a standard of the best interest of the person in question.

Conservatorship and guardianship, for most families, are drastic actions—to be used only as a last resort. Family relationships can be terribly strained when a child alleges that a parent has become incompetent and/or if more than one child wishes to be named conservator or guardian or if no child wants the responsibility.

The court procedure for appointing a guardian or a conservator is expensive and often emotionally trying for loved ones. It's also governed by very restrictive rules. The person appointed will be subject to the jurisdiction of the probate court and be required to make annual reports to the court. It may not be possible, for example, to reduce the size of a taxable estate, which means financial losses.

Many of these problems can be avoided if the person, while still competent, creates a durable power of attorney, a living trust, and/or a living will. Let's discuss those three tools.

Power of Attorney

This is where some people get confused—and this confusion has caused problems when a power of attorney has failed to do what it was intended to do. But it's really not complicated at all.

A power of attorney is a document that a person (the principal) signs in order to give another person (the attorney-in-fact) authority to conduct affairs, make decisions, and carry out tasks. However, there are several variations:

- A general power of attorney grants authority to act on your behalf unconditionally and indefinitely, but the authority ends if you become incapacitated.

- A limited or special power of attorney grants limited authority to accomplish a certain purpose or transaction on your behalf in specific situations or for limited time periods.

- A durable power of attorney grants authority to act on your behalf unconditionally and indefinitely and it remains in effect even if you become incapacitated.
- A springing power of attorney is a variation on the durable power of attorney, but the authority takes effect only if you become incapacitated, and then only after the doctors have made the necessary diagnosis.
- A durable power of attorney for financial affairs grants authority to handle your financial affairs only.
- A durable power of attorney for health care grants authority to make health care decisions only. (We'll discuss this one a little later.)
  - **Power of Attorney:** Written authorization for someone else to conduct affairs, make decisions, and carry out tasks on your behalf.
  - **Attorney-in-Fact:** Person to whom you give the authority to conduct affairs, make decisions, and carry out tasks on your behalf through a power of attorney.

Only the springing and durable powers of attorney provide the protection that the principal needs against incapacitation. The difference is that the springing power of attorney takes effect only when the doctors have diagnosed incompetence, a diagnosis that may come some time after the person in question has lost the ability to make sound decisions. This is why a true durable power of attorney may be preferable.

Making it Durable

A durable power of attorney states that the person whom you authorize to act on your behalf will continue to exercise that authority in the event that you lose the capacity. If a durable power of attorney is properly prepared, the person whom you've designated as your attorney-in-fact will be able to handle all of your financial affairs. It will not be necessary to have a court appoint a guardian or a conservator.

There will be no restrictions. There will be no added expense. You'll have somebody you trust to take care of your estate when you most need help.

A durable power of attorney gives your attorney-in-fact the legal right to assume responsibility for your financial world. He or she can write checks from your account, pay your bills, buy or sell investments for you, sell or give away any of your property, file your income taxes, and so forth. As the title "attorney-in-fact" indicates, the person you choose will be acting as your legal representative.

Whom should you designate as your attorney-in-fact? Somebody you trust, of course, and who knows you, understands your financial affairs, and is capable of doing as you would wish.

If you're married, a logical choice might be your spouse—if he or she is willing and able to assume that responsibility—or an adult son or daughter. Don't name two or more people to share that power or there will likely be problems, although you should also name an alternate.

What makes a power of attorney durable? The document should contain a sentence such as, "This power of attorney shall not be affected by the subsequent disability or incapacity of the principal."

It's usually not necessary to have an attorney draft the document. Most attorneys don't really have any specialized knowledge in this area. In fact, almost all states have a standard or statutory form.

Your attorney-in-fact would have the right to make gifts on your behalf. This right would be useful, for example, if you're confined to a nursing home and you want to protect your assets from being used to cover nursing home expenses. Your attorney-in-fact would be able to carry out your wishes by making gifts to family members, as allowed by Medicaid/Medical Assistance rules. If your estate is large enough to be subject to estate taxes, your attorney-in-fact would be able to make gifts to family members to reduce eventual estate taxes.

If you create a durable power of attorney that authorizes someone to manage your financial affairs, you should put those affairs in writing so that your attorney-in-fact will know all about the nature and extent of them. You probably have most of your affairs in order already if you've prepared a will recently.

Funded Living Trust

Some people choose to plan for possible incapacitation by using a funded living trust. That approach to estate planning is relatively simple in concept, but it can be somewhat complicated in practice.

You set up a trust and transfer title of your assets to the trust, which you will administer as trustee. You name a successor trustee to take over the management of the assets owned by the trust in case you become incapacitated.

> **Funded Living Trust:** A trust that a person creates during his or her lifetime to arrange for management of his or her assets.

But here's where things get complicated. What if you fail to transfer all of your property to the trust? What if you receive income from pensions, Social Security, and other sources? That's when it's valuable to have a durable power of attorney, to take care of all of your assets, current and future, incoming and outgoing.

Your attorney-in-fact can transfer the remaining assets to the trust. The trustee can then manage the assets on your behalf, according to the terms you set out in the trust agreement.

Take a look at your estate plan. Does it include a durable power of attorney and a completely funded trust? If not and you become incapacitated, your family may go through pointless suffering and legal complications to manage your assets.

Durable Power of Attorney for Health Care

A durable power of attorney for health care is a specialized form of power of attorney that authorizes an attorney-in-fact to make health-care decisions for a principal who has become incompetent.

Durable powers of attorney for health care are valid in all states. State laws govern the extent of the power exercised by an attorney-in-fact. Many states have adopted statutory forms, which usually list the limitations placed by law on an attorney-in-fact.

A health-care power of attorney must be signed by the principal or someone acting on behalf of the principal and it must be witnessed by two persons at least 18 years of age or acknowledged by the principal in front of a notary public.

Note: A durable power of attorney for health care is sometimes called a health-care proxy and the attorney-in-fact is then known as the agent.

Just as a legally competent person can create a power of attorney, he or she can revoke it at any time. The only requirement is the principal be legally competent at the time of the revocation.

For the revocation to be effective, the principal must notify the attorney-in-fact (preferably in writing) that the power of attorney is terminated as of a certain date. In the termination notice, the principal should demand that the attorney-in-fact return any power of attorney assets to the principal by that date. The principal should attach a copy of the written termination to the power of attorney document. If the principal has recorded the power of attorney, the revocation must also be recorded.

Choosing a Health-Care Agent

Choosing an attorney-in-fact for health care or a health-care agent is often the most difficult part of this process. You should choose somebody who knows you well, understands your values and beliefs, and can be a strong advocate who will make your wishes known and respected.

A little later in this chapter we provide a list of a few issues that you should discuss with the person you'd like to name to represent you in your durable power of attorney for health care or health-care proxy.

Living Wills

A living will is like a traditional will in only two respects:
- Both wills express your wishes in a formal, written way.
- You must be competent to make either will.

A living will is a document spelling out how much and what kind of medical care the person making the will (the declarant) wants should he or she become terminally ill and incapable of communicating his or her wishes. A person may also designate someone as proxy to make these decisions.

> **Living Will:** A document in which a person specifies the kind and extent of medical care he or she wants in the event that he or she becomes terminally ill and incapable of expressing his or her wishes.
>
> **Declarant:** A person who makes a living will.
>
> **Proxy:** A person designated in the living will to make health-care decisions for the declarant when he or she can no longer do so.

A common misperception about living wills is that they are appropriate only for people who do not want extraordinary measures taken to sustain their life. This perception is incorrect. The treatment choices may range from none at all to every possible means of sustaining life.

You do not need a lawyer to draft a living will, although many people have them drafted at the same time that they are having a traditional will drafted. Many people also seek advice from a doctor or religious advisor before drafting a living will because most people would not be able to describe their wishes specifically without first researching the kinds of medical technology currently available to them.

A living will can be very detailed:

- You can direct that certain treatments be given for specific illnesses.
- You can specify a preference for home, hospital, or hospice treatment.
- You can make known any religious objections to a particular treatment.
- You can identify, in advance, any individuals who are likely to try to interfere with treatment decisions and clarify your wishes with regard to those persons.

The most important point about a living will is that it allows you to make decisions in advance about how much and what kind of health care you want. Since the purpose of a living will is to serve as a guide to those who need to make decisions about your care, the more you detail your wishes, the more helpful it will be.

In some states, the law allows a living will to express wishes concerning organ donation. You may state whether you wish to donate your organs upon death and specify any limitations or special wishes.

For a living will to be valid, there must be at least two witnesses or one notary public. A living will becomes effective when it's delivered to an appropriate health care professional. You should also give a copy to a family member and any family clergy.

It's important for your doctor to know and understand your wishes. When you give your living will to your doctor, he or she must tell you whether or not he or she will comply with it. A doctor has a right to decline to follow the terms of a living will, within the limits of reasonable medical practice. If that's the decision, he or she must help you find another doctor and must make sure that the new doctor is aware of the living will.

Finally, a living will is revocable. You may revoke your living will totally or partially at any time in any manner, regardless of your physical or mental condition. States vary in what they call a living will and a durable power of attorney. These forms are often called "advance directives." (In fact, some states have combined living wills and health-care power of attorney into a single advanced directive document.) States also have different requirements for making the documents legal.

You can generally obtain state forms and literature from:

- Your local hospital
- Your local nursing home
- Your state or local office on aging
- Your state's bar association
- Your state's hospital association
- Your state's medical association

You can also get advance directive forms from Choice in Dying, a national nonprofit organization. Choice in Dying provides, for a nominal fee, advance directive forms tailored to each state's legal requirements. It pioneered living wills in 1967 and has distributed more than 10 million advance directives since then. The organization monitors legislative changes nationwide and updates all state documents accordingly.

Health Care: Choices, Beliefs, Values, and Feelings

How does your faith community, church, or synagogue view the role of prayer or religious sacraments in an illness? What else do you feel is important for your agent to know?

Inform your health care agent about any changes in your health or in your beliefs or attitudes. How well your health care agent performs depends on how well you have prepared him or her.

As a health-care consumer, you have rights and choices regarding health care. Because making health-care choices may raise some sensitive issues, it's often a difficult subject to discuss.

The purpose of this section is to help you think about what could happen to you and to help you choose the most appropriate care based on your personal needs and beliefs. As a way to guide you through this process, we've included an explanation of commonly used medical terms and questions to help you reflect on your values.

A national law, the Patient Self-Determination Act, went into effect in December 1991 that set universal standards for informing patients of their legal options for refusing or accepting medical treatment. The bill affects all health-care facilities that accept federal funding—hospitals, health maintenance organizations, home health-care services, nursing homes, and hospices.

The law requires each of these facilities, when admitting a patient, to provide a form explaining the state's law regarding acceptance or refusal of medical treatment and the institution's policy concerning it. The staff must ask if you have a living will or health-care proxy and, if not, to make certain choices about your future treatment, in a declaration to physicians.

This is your choice. You are not required, under any circumstances, to create a living will. In fact, it's against the law to require anyone to write a declaration to physicians.

> **Declaration to Physicians:** A form of living will, a document that specifies your wishes concerning health-care treatment.

But if you choose to create one, it must be respected. If you are unable to participate in decision making, in most cases the physician will consult with your family and then your living will.

It's not necessary to complete the form you receive from the health-care facility, but you must answer the question about artificially administered sustenance. Again, you may state that you want your proxy to make those decisions.

Living wills are interpreted only by a doctor. They do not apply in home emergency situations. Paramedics will provide full emergency care.

Once you have written a declaration to physicians or a living will, review it periodically to make sure it continues to be a current statement of your wishes. If you revise it, it's best to fill out a new declaration to physicians and distribute it to all appropriate persons and institutions.

> **Artificially Administered Sustenance:** Providing special nutritional formulas, fluids, and/or medications through tubes when a person cannot take food or fluids by mouth.

Explanation of Medical Terms

Terminal Condition: This is an incurable or irreversible condition in which any medical treatment will only prolong the dying process.

Vegetative State: This is a state in which a person is unable to talk or think or understand others. This condition can be caused by strokes and other diseases of the brain. It is irreversible except in rare cases.

Life Tenancy

We discussed life tenancy earlier, in the context of setting up a life estate, a type of ownership that avoids probate. As you recall, the way it works is that you transfer property to your children, but you reserve the right to use the property until you die. In legal terms, you are the life tenant and your children are the remaindermen.

We should point out ten possible adverse effects of this type of transfer:

1. If you wish to sell or mortgage your property, it may be awkward because all of your children and their spouses must sign the deed or mortgage.

2. If any of your children have a judgment or tax lien, it may well attach to their remainder interest. This will usually mean that it must be satisfied before the property can be sold or mortgaged, resulting in a loss to your child. If a child later develops financial problems and files for bankruptcy, he or she will lose the remainder interest.

3. If any of your children have marital problems that end in divorce, remainder interest may figure in the property settlement and may pose a problem.

4. If the property is sold before you die, if there's a taxable gain, your children will have to pay income tax on a portion of the gain.

5. If the property is sold during your lifetime, your children will receive part of the sale proceeds and they will have no legal obligation to return any portion of it to you.

6. If you are receiving nursing care and the property is sold, a portion of the sale proceeds will be used to pay for your nursing care expense. The percentage is determined by published actuarial tables.

7. If any of your children die before you, it will be necessary to probate that child's interest. Usually, a remainder interest owned by a deceased child will go to his or her spouse. It will

then require some special effort to have the surviving spouse transfer the property to grandchildren.

8. The property will be included in your taxable estate for estate tax purposes.

9. In the event of a sale of just your life estate or a sale by just one of your remaindermen, the term interest rule will apply. That means that no income tax basis is allowed on the sale and the entire sale price is treated as a taxable gain.

10. If, at a later time, you want your children to give back their remainder interest, the gift back will be regarded as a future interest and, therefore, part of their unified credit will be used up.

Joint Tenancy Issues

Jointly owned property is probably the least understood area of estate planning. People often refer to joint tenancy as the poor man's will. It can be useful in estate planning to put your property into joint tenancy with your spouse, child, parent, or another family member—if you know what you're doing.

Joint tenancy with right of survivorship means that each joint tenant has a full and undivided interest in the property. (The joint tenants are usually husband and wife, although they could also be business partners.) Bank accounts and real estate are the types of property most often held in joint tenancy.

Neither party can sell the property without the other's consent. The only exception to the rule is that either joint tenant can withdraw funds from a joint bank account. Upon death of a joint tenant, the entire property passes to the survivor(s) automatically, avoiding probate and the courts.

At least one advantage to owning property in a joint tenancy is that it avoids probate. But there are certain disadvantages to owning property in joint tenancy:

Unaffected by Will. No joint tenant can transfer joint property by means of will instruction. A will distributes only property in an individual's name or an interest held in tenants-in-common. Since joint property transfers by right of survivorship, you cannot control the disposition of joint property by will!

Undesired Beneficiaries. Joint property may pass to someone you don't wish to receive it. Let's assume that a husband and a wife own all of their property jointly. If the husband dies first, his estate will automatically pass to his wife. Now the wife controls all of the property, and she can give it upon her death to whomever she wants—a new spouse, for example, and that spouse's children.

Incompetent Spouse. The surviving spouse may not be experienced in money matters, or may be physically or mentally disabled. If this is the case, the survivor may well end up in probate for a living probate proceedings to determine who will act as conservator for that surviving spouse.

Estate Tax Issues. Since all joint property goes to the surviving spouse, it may be subject to estate tax when the surviving spouse dies. This is not a problem unless the surviving spouse's estate is larger than the amount of the unified credit/personal estate tax exemption. For 2005, the personal estate exemption is $1,500,000. However, it's very common with large estates for a husband and wife to do estate planning to minimize estate taxes, but forget to change the joint titling of their assets. This is done by having special provisions written into a will or trust that implements the credit shelter of the first spouse to die, thereby utilizing the personal estate exemption of both spouses. With joint tenancy ownership it does not work!

Income Tax Issues. Joint tenancy may create an income tax problem if the estate owns appreciated assets. If an individual purchased a parcel of real estate for $50,000 and at death it's worth $200,000, the surviving spouse who receives the real estate by means of a will can sell it for $200,000 and pay no capital gains tax. The income tax cost basis of the real estate is stepped up at death to the current market value that the real estate was valued at in the estate.

If the property was received through joint tenancy at death, the surviving joint tenant would be taxed on capital gains of $75,000. The basis in the property would be one-half of the purchase price of $50,000 ($25,000) plus a stepped-up basis at death on the other one-half of the property ($100,000), which would equal $125,000. The selling price of $200,000 would create a $75,000 capital gain.

Unwanted Family Trouble. If an individual who transfers an asset to joint tenancy later decides that he or she wants it back, unwanted family tensions are often created. After the transfer of assets, the new joint tenant may not want to relinquish ownership in the asset. This could happen in a situation where a son or daughter is placed as joint tenant on a parent's asset. This situation can cause a stalemate and create problems that could well have been avoided. Jointly held bank accounts may be the most common cause of problems.

Many single people place bank accounts in joint tenancy with one or more of their children. Many elders feel that this is a simple strategy that allows a son or daughter to receive the money after death without any probate. However, if the child gets into a financial trouble (bankruptcy, divorce, or lawsuit), it may well prove disastrous for the parent, because one half of the asset now belongs to that child and may be pursued by creditors.

There are better strategies, such as the Totten trust, annuities, and life insurance. You can also sidestep probate by establishing a living trust.

Joint tenancy is still the most common form of family ownership in America. It performs an important role in small estates. However, if you've accumulated wealth in excess of the personal estate exemption or you own and operate a business, you should develop a more advanced strategy in your estate plan.

8

Getting More Out of Giving—
Charities and Your Estate

Charity means giving from your heart. Now we're going to show you how you can also give from your head. In this chapter we'll try to cover some of the basics and provide enough information to highlight the benefits of charitable giving in estate planning.

Many people give to charities, regularly or at least occasionally. They give to religious organizations, educational institutions, hospitals, local and national charities, and foundations.

The contributions that charities receive from generous donors help keep them operating. In fact, without that generosity many charities would cease to operate.

How many people with good intentions would give even more if they knew how? There are ways to give from the heart and from the head. That's where estate planning can help all of us help our favorite charities and get more out of giving.

Will

Probably the most common way to benefit charities—and reduce your taxable estate—is to make a charitable bequest through your will. Any gift bequeathed to an approved charity is exempt from federal and almost all state gift and estate taxes. That tax advantage provides a little extra incentive to give.

But which charitable causes should you support with a bequest? Maybe you already have some favorites. If not, then we recommend that you contact the following agency:

National Charities Information Bureau
19 Union Square West
New York, NY 10003-1997
Phone: (212) 929-6300
Fax: (212) 463-7083
Web: **www.give.org**

The NCIB publishes the *Wise Giving Guide* quarterly, which includes a Quick Reference Guide that lists close to 400 national organizations. The guide can help you make a more informed decision about donating.

It's relatively easy to name a charity as a beneficiary of your will. Many charities, especially churches, conduct seminars to explain how to do just that. But there are two points that we'll mention here because they're important and can avoid problems for your estate.

The first point is to make sure that whatever charitable cause you choose qualifies as tax-exempt. Normally, this means that the organization has tax-exempt status under Section 501 (c) (3) of the Internal Revenue Code. This status should be stated somewhere on any materials provided by the organization. If you're not sure about an organization you're considering, contact the IRS to check whether it's on the list.

Note: The IRS has installed an electronic version of Publication 78, "Cumulative List of Organizations," on its Website: http://www.irs.ustreas.gov/prod/bus_info/eo/eosearch.html. Just enter the name and city and state of an organization and you get a brief statement about it, including the percentage of the deductibility limitation.

The second point is to have a backup plan. It's not likely to happen, but what if an organization you name as beneficiary loses its tax-exempt status or just ceases to exist? In the former case, your bequest will still go to the organization, but you'll lose any tax advantages. In the latter case, your bequest will probably end up in your residuary estate.

That's where it's wise to have a backup organization. You simply name a second beneficiary, just as you would for any other beneficiary of your will, as we explained in Chapter 4.

Planning for Leverage

Contributing to charities through a will is a very good strategy. But it has no impact on your financial position during your lifetime. Yet many people would like to do more while they're alive—particularly if they knew how to leverage their dollars for the benefit of the charity.

In many cases, charitable planning shifts money from taxes to charity. (If your favorite charity is Uncle Sam, the next few pages may be of little interest to you.) Planning can allow you to support charitable causes yet maintain or even increase the amount you leave to your heirs. We'll focus here on two basic types of giving: Life insurance and charitable remainder trusts.

Life Insurance

A very simple way to leverage a gift to your favorite charity is through life insurance. You just name the charity as the beneficiary and as the owner.

Because life insurance is an investment, it generally pays out significantly more than you pay into it. This appreciation in value makes it an excellent way to leverage your charitable dollar.

When you make a charity the owner and beneficiary of your life insurance policy, you gain two advantages. First, you can deduct the premiums from your income taxes as a charitable contribution. Second, your gift removes the policy proceeds from your taxable estate.

Transfer of Assets

Pop quiz: Which is worth more, $100,000 in cash, or real estate that has appreciated in value from $25,000 to $100,000?

Answer: The property is worth more—at least if you give it away.

Let's take an example. Assume that John is planning his estate and wants to make a charitable contribution of $100,000. John owns a piece of real estate that he purchased for $25,000 and that is now worth $100,000. He plans on selling the property because he does not need it and believes it has peaked in value. John also has a discretionary $100,000 in cash to give to the charity. Which asset should he give?

If John gives the cash, he gets an income tax deduction of $100,000. If John gives the real estate instead, he still gets a $100,000 deduction, but he escapes the capital gains tax that he would have faced if he'd sold the property.

What does John gain in this transaction? He receives a charitable deduction of $100,000 and saves the tax he would have paid on $75,000 of realized capital gain. Assuming that his capital gains tax rate is 15%, he saves $11,250 in tax.

(The benefit to the charity remains the same whether John donates cash or property. So, if you answered the pop quiz with "Neither," you're right—from the perspective of the charity. But John would have another perspective.)

Let's assume now that John wants to benefit the charity but is unwilling to give away $100,000. Also, he wants to sell the real estate and reinvest the net proceeds to receive income, but he's reluctant to sell because of the capital gains tax.

There's a way that John can get what he wants and at the same time benefit the charity. It's called the charitable remainder trust.

Charitable Remainder Trust

What's a charitable remainder trust? A CRT is a trust that benefits both you and whichever charity you choose.

GETTING MORE OUT OF GIVING—CHARITIES AND YOUR ESTATE

How does it work? It's as easy as 1-2-3:

1. You set up a trust with a one-time contribution of principal.
2. You name two types of beneficiary: An income beneficiary (yourself, someone else, or several people) and a final beneficiary (the charity of your choice).
3. You designate a period of time, or life.

The trust then pays an annual sum to you or whomever you name as income beneficiaries over a specified period. Then when the income beneficiary dies, the charity named as final beneficiary receives the remainder of the assets in the CRT. (That's why it's called a remainder trust.)

That's easy enough, even when we throw in some numbers. The annual payments to the beneficiaries must be at least 5% but no more than 12%. The present value of the amount going to the qualified charity must be at least 10% of the amount contributed. When that specified period ends, the remainder of the principal is transferred to the charity.

> **Charitable Remainder Trust (CRT):** A gift made in trust to a qualified charity, an arrangement that regularly pays income from the assets to the donor or another beneficiary during the donor's lifetime and then passes the remaining assets to the designated charity.

(We should remind you that if the income beneficiary is anyone other than you or your spouse and the annual payout is more than $10,000, you'll have to pay a gift tax. So don't be too generous with the payout—or split it among two or more beneficiaries.)

OK. Now let's throw in another detail. The charitable remainder trust is available in two forms: The annuity trust and the unitrust. The difference is slight, but potentially significant:

- The annuity trust pays the income beneficiary a fixed amount each year, no matter what—a steady income stream.
- The unitrust pays a percentage of the value of the trust assets and any accumulated earnings, as evaluated every year. If the unitrust assets appreciate, the payout increases; if the assets lose value, the payout is less.

Whether the CRT is set up as an annuity or as a unitrust, it's a trust that works partly for the benefit of the donor (in most cases) and partly for the benefit of the charity.

Let's use as an example the situation with John and his $100,000 piece of real estate. John transfers the real estate to a charitable remainder trust. The trustee then sells the real estate and reinvests the proceeds of $100,000. There's no capital gains tax because the CRT is a tax-exempt entity. Then, every year John receives income from the trust. If he chooses the minimum 5% payout rate, he receives $5,000 exactly (if it's an annuity), or more or less (if it's a unitrust).

John pays income taxes on the income as he receives it. But he would have paid income tax up front if he'd sold the real estate and reinvested the difference. If he had sold it, he would have had only $85,000 to invest, which at 5% would have yielded $4,250.

Figure 8A shows how John uses the CRT to receive an additional $750 a year, transfer the property for the benefit of the charity, and remove the $100,000 from his estate without incurring a gift tax or using any of his unified credit.

Wealth Replacement Trust

When we discuss transferring an appreciated asset transferred to a charitable remainder trust, people frequently ask, "What about my heirs?" That's a very good question. And we have a very good answer—the wealth replacement trust. This trust does exactly what the name states: It compensates the heirs for the assets transferred to a charitable cause. The wealth replacement trust is an irrevocable life insurance trust (ILIT) that a person sets up as both owner and beneficiary of a life insurance policy on his or her life, as we discussed in Chapter 6.

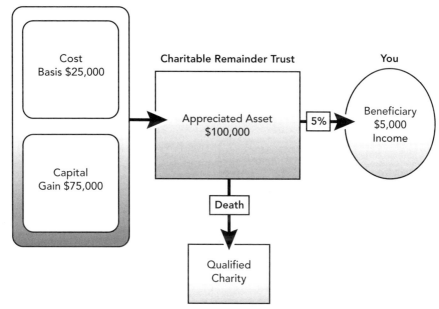

Figure 8A: CRT Trust

The face value of the policy will be the amount the donor wants to leave to the heirs—usually the value of the asset given to the charitable remainder trust. The donor pays the premiums on the life insurance policy each year.

(One source of funds for paying these premiums may be the savings from the charitable contribution.)

The trustee of the wealth replacement trust (this is often the donor) then pays the premium due on the life insurance policy out of the funds transferred to the trust. This happens every year until the donor dies or pays in full for the policy. Then, when the donor dies, the proceeds from the life insurance policy are distributed to the heirs as beneficiaries of the trust. As we discussed, the ILIT holds the policy outside the estate, so the proceeds are not subject to estate tax.

> **Wealth Replacement Trust:** A trust set up to compensate heirs for a contribution to charity of assets that would otherwise have been included in the estate for the heirs.

Of course, the IRS places a few restrictions on wealth replacement trusts. The tax code requires that the heirs have a present interest in the trust. How can heirs have a "present interest" in a life insurance policy? By having the right to the cash paid for the premiums. So, every year, before gifting the premiums into the trust, the donor must inform the heirs that they have a right to the money and they must then waive this right.

Let's look at the example of John's transfer. Obviously, after he sets up a CRT for his $100,000 property, his estate will be worth $100,000 less to his heirs. If that's a problem, here's a solution for John and maybe for you, too.

> **Present Interest:** The right to use a gift immediately.

As **Figure 8B** shows, John transfers $100,000 of appreciated property to a CRT. Then, to replace the "lost" assets in his estate, John purchases a life insurance contract inside an irrevocable life insurance trust. The ILIT holds the policy outside the estate for estate tax purposes and John has an additional $750 annually that he can use to pay the insurance premiums. When John dies, the assets in the charitable remainder trust pass to the charity and the $100,000 death benefit from the life insurance inside his wealth replacement trust passes to his heirs without estate tax.

The significance of this becomes clear when we analyze the combined savings in capital gains tax, income tax, and estate tax. First, John saved approximately $11,250 in capital gains taxes by transferring the real estate to the CRT. Second, he receives a charitable income tax deduction in the year in which he places the property in the trust. (The amount depends on his age.) Third, exclusion of the real estate from his taxable estate saved $41,000 in federal estate taxes.

Figure 8B: Wealth Replacement Trust

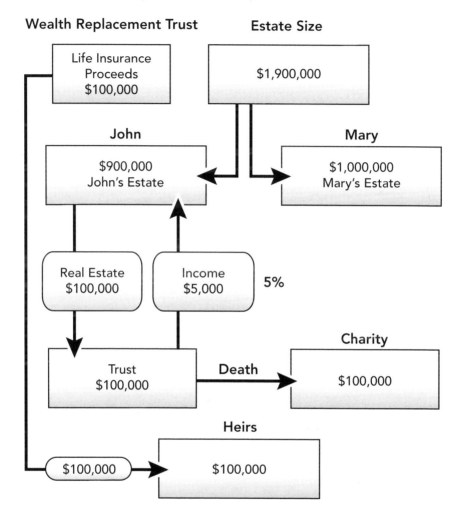

The Bottom Line

John saves at least $52,250 in taxes. By using excess earnings to help pay premiums on the $100,000 of life insurance, John replaces the asset to his heirs at no cost.

That's a creative way to use a CRT. Here's another, which can solve a common problem in taxable estates—retirement plans.

Retirement plan distributions are subject to income tax and the assets remaining in the plan at death are subject to estate tax. If the account is substantial, that can cause a problem.

So how do you avoid those taxes? You set up a charitable remainder trust as the beneficiary for your retirement plan and name your spouse as income beneficiary of the CRT.

As we see in **Figure 8C**, John has $200,000 in a qualified retirement plan. He names as beneficiary a charitable remainder trust. This strategy removes the $200,000 from his taxable estate. He sets up the CRT to pay income to Mary and then, when she dies, the $200,000 in the trust passes to the designated charity.

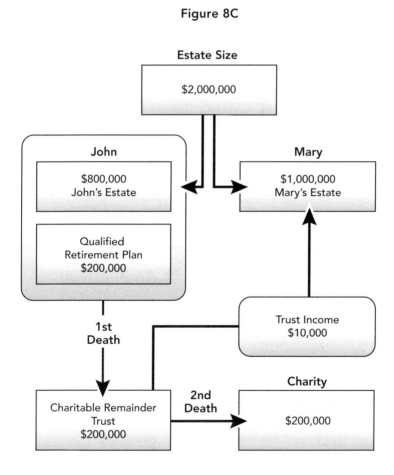

Figure 8C

Here's the result. There are no income or estate taxes payable at death because qualified charities don't pay income or estate taxes! The retirement plan is removed from your taxable estate. Your spouse pays income tax on the distributions that are made, and if he or she doesn't need the income, it can be used to fund a wealth replacement trust for the children.

Note: If you want to use this strategy for taking care of a qualified retirement plan, you'll need special legal documents. We recommend that you consult a tax attorney to handle this matter.

Leaving your qualified plan assets to a specific charity or to a charitable remainder trust will save estate and income tax. By giving your qualified plan assets to charity, you create a 100% charitable deduction for estate and income tax purposes. If you use a CRT, the charitable deduction is based on the present value of the remainder interest left to charity.

You can transfer assets to a charity either during your life or at death. In both cases, the assets are removed from your estate without using any of your unified credit on the transfer. Generally, a transfer to charity or a charitable remainder trust does not create an income tax or estate tax liability. It does not act as a trigger for capital gains on appreciated assets and it may even create an income tax deduction.

So, give until it feels good, emotionally and financially.

＃ 9

The Family-Owned Business
Structures, Strategies, and Succession

All of the chapters up to this point have focused on the individual, on personal financial matters in general. In this chapter, we turn to the particular concerns involved in owning a business.

As you know, a business can be owned in various ways, from a small sole proprietorship to a large publicly owned corporation. Each structure has its own unique inherent risks and rewards.

If you own a business, you're well aware of the risks of operating a business—competition, shortages of materials or labor, excess or inadequate production capacity, loss of important clients or customers, dramatic fluctuations in cash flow and prices, and changing economic conditions.

However, you may not be aware of the even greater risks that come with the form of the business you own. Whether your business is a sole proprietorship, a partnership, a closely held corporation, or a professional corporation, its structure poses certain risks that threaten the survival of your business and perhaps even the financial future of your family. We'll look at each type of business organization and discuss the risks that arise from the way each is structured.

Sole Proprietorships

A sole proprietorship is the easiest form of business to organize and to operate. There are no special legal requirements for either starting up or terminating a sole proprietorship. Because of the ease of starting and running this type of business, sole proprietorships are the most common form of business in the United States.

In addition to its simplicity, a sole proprietorship provides the flexibility to change the focus or the direction of a business. Since the business is owned by one individual, the owner is free in making business decisions.

The owner can determine what will happen to the business at retirement and, with proper planning, at death. The assets used in the business remain the personal property of the proprietor or his or her estate, instead of being titled in the name of the business, which would make it difficult for the business owner or heirs to get them back.

The Risks of Sole Proprietorship

The greatest risk that confronts a sole proprietorship may be the death of the proprietor. When the proprietor dies, in almost all cases the business dies as well. The business that has been the sole or primary source of income for the owner and family is worth little or nothing after the owner dies, except for any assets. The deceased owner's family is usually forced to liquidate the business at less than its actual value to pay estate settlement costs and taxes.

Since a sole proprietor's personal and business assets are usually one and the same, they both can be subject to Attachment by the creditors of the business if the business gets into financial difficulty. The owner could lose all assets, personal as well as business. Sole proprietorships also face problems of raising additional capital, since they can't bring in outside investors through issuing stocks and bonds. The proprietor must rely on funds generated by the business or borrowed from the bank, which inhibits the growth of the business. Selling or transferring a sole proprietorship can be difficult because the assets of the business usually have to be sold separately.

And when the proprietor dies, the executor of the estate must take possession of personal and business assets and pay personal and business debts.

Partnerships

If two or more individuals find the sole proprietorship inadequate for the needs of their business but they don't want to go to the expense of setting up a corporation, the partnership may be the ideal compromise solution. A partnership is simply the association of two or more people who have agreed either verbally or in writing to combine their skills and resources for profit.

In a general partnership, each partner contributes a share of the capital and services to the business and has a share in the control, management, and liability of the business. Each partner is a principal in the business and has equal authority with the other partners and shares equally in the profits or losses, assuming that the partners have all made equal contributions to the business and they have not modified that equality by written agreement.

The Risks of Partnerships

The greatest risk of running a business as a general partnership is death: The death of a partner may dissolve a partnership automatically and instantaneously. The law requires that a partnership be dissolved upon the death of a 50% general partner—even if the partners have prepared agreements to the contrary.

The surviving partners could continue the business with or without a representative from the deceased partner's estate or family, but they would have to reorganize in the form of a new partnership. And the heirs of the deceased partner may be legally entitled to demand liquidation of the firm and division of the assets, if they choose.

Another risk in the general partnership form of business is the unlimited liability of all general partners for the debts of the partnership and any negligent acts of the other partners and employees. Each partner's personal assets are at risk for payment of the partnership liabilities.

Corporations

The corporate form of business grew out of a need to provide business with more stability and continuity than is found with the simpler forms of business organization. Although the variety of corporations has expanded to include closely held corporations, publicly owned corporations, C corporations, S corporations, and Professional Corporations, they all share the following features:

- A corporation is a separate legal existence, independent of its owners, which means the business does not have to be dissolved upon the death of an owner.
- Ownership is evidenced by possession of stock, which entitles the owner to a proportionate share of profits and provides a vehicle for transferring ownership.
- The liability of each owner is limited to the amount of his or her investment.
- Management of daily affairs is by officers selected by the board of directors, who are elected by shareholders.
- Profits are distributed as dividends to stockholders.

Although all corporations share these characteristics, they can vary with respect to ownership, tax status, and the type of business conducted. These are the basic types of corporations:

C Corporations: These are conventional corporations that pay tax directly to the IRS. They can be publicly owned or closely held. They are named C corporations after Subchapter C of the Internal Revenue Code.

S Corporations: These are corporations that have elected to be taxed like partnerships. The corporation generally pays no tax; instead, all income and losses pass through directly to the stockholders, who pay taxes on their shares. These are named S corporations after Subchapter S of the Internal Revenue Code.

Professional Corporations: These are corporations composed exclusively of professional service providers, such as doctors, lawyers, accountants, architects, and others licensed to practice a "learned profession" or provide a service. These corporations must file articles of incorporation with the state that meet the state's specific requirements.

Closely Held Corporation: A corporation whose shares (or at least voting shares) are held by a single shareholder or closely knit group of shareholders, who are generally active in the conduct of the business, which usually has no public investors. Also known as a close corporation.

The Risks of Incorporation

Incorporation eliminates or minimizes many of the risks inherent in a sole proprietorship or a partnership, such as liability for business obligations or the dissolution of a business at the death of an owner. However, it doesn't make the owners of a corporation—especially one that is closely held—immune to all misfortune.

The death of a major stockholder in a corporation can jeopardize the future of the business as well as the income and financial security of other shareholders.

When a stockholder in a closely held corporation dies without a business continuation plan, that situation naturally divides the shareholders into two groups with opposing priorities. On the one hand, the heirs and executor of the deceased shareholder will be concerned with immediate needs for income to pay estate settlement costs and provide for the family's financial needs. On the other hand, the surviving shareholders will be more concerned about long-term future and financial needs of the business.

The potential conflict arising from these opposing priorities is usually resolved by one of the following four alternative courses of action, each of which creates its own risks.

1. **The surviving stockholders buy out the heirs of the deceased stockholder.** While this is the most logical solution to the problem, it can be the source of numerous problems. For example, will both parties be able to agree on a fair price for the deceased shareholder's stock? Can the surviving stockholders buy the stock immediately with cash or will they have to buy it on installment? In the latter case, are the heirs willing to wait?

2. **The heirs enter the business as active stockholders.** Usually the heirs will be the decedent's spouse and children, who may or may not have the necessary experience to manage or significantly contribute to the success of the business. If the heirs are majority stockholders and lack knowledge of the business but insist on managing it, the future of the corporation may be in jeopardy.

3. **The heirs enter the business as inactive stockholders.** Although potentially less damaging than the previous alternative, this option poses its own special problems. Since the heirs will need a source of income to replace that provided by the decedent, they may agree to stay out of the management of the business for a share of the profits. This idea may not appeal to the surviving stockholders, who then have to assume all responsibility for managing the company and pay out a large part of corporate profits as dividends to heirs who contribute nothing.

4. **The heirs sell their stock to an outsider.** This option again raises the issue of determining a fair price for the stock, assuming a buyer can be found. The surviving stockholders

may have no control over to whom the stock can be sold or under what conditions it can be sold. If the deceased stockholder was a majority stockholder, selling his or her stock to an outsider would pass control of the business to a stranger, which could make the surviving stockholders' positions nearly untenable. What's to keep a leading competitor from buying the heirs' stock and taking over the business?

The essence of a corporation also can be put at risk by the departure or death of a key executive or employee, whether he or she is or is not a shareholder. Numerous studies have shown that the primary cause of business failure is a breakdown in management because of the loss of an experienced manager or key person. The loss of a key employee could disrupt the management of a corporation, reduce its earnings, affect the availability of credit, threaten customer relationships, create problems in finding a replacement, and/or jeopardize the loyalty and morale of other employees.

The risk that can result from the death or departure of a stockholder or key employee should make it obvious that the corporate form of business alone is no guarantee of stability and security for stockholders, regardless of whether they have minority or majority holdings.

The most valuable asset of any business is people. We've discussed the potentially devastating impact that the death or loss of an owner, partner, or other key person can have on a business—proprietorship, partnership, or corporation. The risks posed by a loss of control or the loss of a key employee can be minimized or even eliminated by advance planning using well-established business strategies.

BUSINESS CONTINUATION STRATEGIES

The operation of a business can be adversely affected by the death, disability, or departure of a co-owner or of a key person associated with the business. You can't prevent deaths, disabilities, or departures, but you can make plans to ensure that the business will continue.

Cross-Purchase Plans or Buy/Sell Agreements

A cross-purchase plan may be the ideal solution for the problem of retaining control of your business after the death of a partner or co-owner. A cross-purchase or buy/sell agreement is a plan that provides for an orderly change of ownership when a business owner dies or becomes disabled.

This agreement is a legal contract signed by all of the owners of a business. If one of them dies, the others will buy the deceased's interests, which the estate must sell, at a specified price.

> **Cross-Purchase Agreement:** A contract that provides for an orderly transfer of ownership interests (stock or partnership interests) in a closely held business when an owner dies or becomes disabled. Also known as a buy/sell agreement.

In addition, each owner agrees not to dispose of his or her interest in the business without first offering it to the other owners at the previously agreed, upon price.

The agreement serves the purposes of both the surviving business owners and the family of the deceased owner. It enables the remaining owners to acquire the interests of the late owner and ensure continuity of the business, while it provides the heirs of the deceased owner with fair and full payment for their interest in the business.

The only question is how the cross-purchase agreement should be funded. The surviving owners have four alternatives for funding their purchase of the interest:

1. **Pay Cash.** If the surviving owners have adequate personal assets outside the business, they can use accumulated capital for the purchase. However, if these assets are not liquid, the sudden forced transfer could result in substantial losses to the owners. Furthermore, using cash for this purchase will take away funds that could be used more profitably for business purposes.

2. **Borrow the Money.** The surviving owners may find it difficult, if not impossible, to borrow money after the loss of a key

person because of questions about the viability of the business. Even if they succeed in borrowing the money, they would be mortgaging the future income of the business at the expense of profits and working capital.

3. **Create a Special Savings Fund.** The owners could make periodic deposits to a savings fund to accumulate the purchase price. However, one of the owners may die before the funding is completed, so there would still be a liquidity problem.

4. **Use Life Insurance.** Each owner of the business could purchase life insurance policies on the other owners with face values equal to their interests in the business. In contrast to the disadvantages and uncertainties of the other methods, life insurance provides certainty and guarantees the effectiveness of the buy/sell agreement. By using life insurance to pay the purchase price, they pay out discounted dollars and will receive a federal income tax exclusion on the net gain when an owner's death matures the policy. (You may recall our discussion in Chapter 5 of using life insurance to fund estate liquidity.)

Stock Redemption (Entity) Plans

A cross-purchase plan can be impractical for a business with more than two or three owners, because of the number of insurance policies that would be necessary to cover all of the owners. A business with three owners would require six policies, for example, and a company with four owners would necessitate twelve policies.

Rather than buy that many policies, you could set up a stock redemption or entity plan. Under this plan, the corporation or partnership purchases the life insurance on each of the individual owners, becoming the policy owner and beneficiary.

Each policy would have a face value equal to the business interest of the insured, to ensure that adequate funds will be available to purchase the stock of any owner who might die. With such a plan, a business with four owners, for example, would need to purchase only four policies, rather than the twelve that would be required under a cross-purchase plan.

The Advantages of Buy/Sell Plans

There are benefits to buy/sell plans for everybody involved directly or indirectly with ownership of the business.

For the owners while living:
- A buy/sell plan establishes a guaranteed price and market for their interest in the business.
- They can feel assured that their heirs will receive full value.
- Cash values of the policy can provide funds for emergencies, opportunities, and retirement.

For the surviving owners:
- They retain control of the business.
- Creditors and employees are assured that the business will continue.
- Funds are guaranteed and provided when needed.
- Business credit is maintained and even strengthened.

For the heirs:
- They receive an immediate, fair price in cash.
- The estate can be settled promptly and efficiently.
- They are not burdened with business responsibilities.

Family Limited Partnership

Another type of ownership that has gained popularity is the family limited partnership.

A partnership could include parents, children, and grandchildren. The partners also could be entities such as trusts created for the benefit of the family members.

Forming a family limited partnership begins with an accurate assessment of the business and its value. Then there's paperwork (of course!) and a few critical decisions.

When organizing a family limited partnership, the most important decision for the family members may be whom to name as the general partner(s). The general partner(s) will have control over the partnership's business activities and determine how much of the partnership income is to be distributed to the partners.

Also, the general partner has unlimited liability in a family limited partnership. Newer forms of partnerships for tax purposes, limited liability companies, and limited liability partnerships may also allow a managing partner to have limited liability.

The family limited partnership can be an important tool when developing an estate preservation plan. Typically the partnership allows structured management of business and investments. It also protects the family's assets from debt and creditors. The general partner can retain control of assets even after the transfer of ownership.

Often, a family limited partnership is established to allow the transfer of limited partnership interests from parents to children without transferring control. A family limited partnership ensures continuous ownership of a family business.

It also allows potential discounts for federal estate and gift tax on transfers of limited partnership interests. An estate planning strategy is to have the majority interest, owned by the individual with the highest net worth and his or her spouse, gradually gifted so that it is converted to a minority interest subject to discounts.

Discounts can be taken from the family limited partnership's value when minority and unmarketable interests are gifted to descendants. These discounts further reduce the value of the estate for estate and gift tax purposes.

Various discounts can be created:
- A lack of control discount is available for minority limited partnership interests. Because limited partners are not able to influence management decisions, the value of their interest is discounted to reflect this lack of control.

- A lack of marketability discount is available and reduces the value of privately held limited partnership interests that do not have a market for trading.

Situations in which a family limited partnership may be appropriate:
- To shift the income from a parent in a high tax bracket to a child in a lower tax bracket.
- To protect assets after a parent transfers partnership interests to younger generations when they otherwise may be lost due to poor management or divorce.
- To conduct a family business in a form other than a sole proprietorship.
- To provide flexibility in establishing rules for managing property.
- To simplify ownership and gifting of assets.
- To ease the distribution of assets at death among family members without having to remove the assets from the partnership.

Implementation of a succession plan for management and estate planning purposes is also very important. It can be achieved by choosing a non-managing general partner who will take over management after the general partner's interest is terminated.

LIMITED LIABILITY CORPORATIONS

A limited liability corporation is designed to attain one of the benefits enjoyed by stockholders of corporations, namely limited liability. Asset protection is a benefit of this form of operation. Also, the LLC has little ownership and operational restrictions.

Shareholders in an LLC are responsible for debts only up to their individual investment.

The creditor, however, can request personal guarantees. The tax treatment and distribution rules of the limited liability corporation are complex and beyond the scope of our discussion here.

A limited liability company (LLC), however, is taxed as a partnership. Losses incurred by the LLC pass through to owners (members) and are deductible under partnership rules.

An operating agreement that governs management is the fundamental element of the LLC. LLC voting privileges in most cases are directly related to the capital contributions of a member. Occasionally, with careful drafting, it also can be based on profit participation.

LLCs differ from limited partnerships in several ways. Limited partners of a limited partnership do not participate in management decisions. Members in a limited liability company may participate in management. General partners of a limited partnership participate in management and are liable for the debts.

The transfer of family business assets to family limited partnerships has become a common estate planning technique. An LLC can serve this purpose as effectively. Additionally, no family member has to perform as the general partner and assume liability or debts of the business.

Special Land Use Valuation: Section 2032a

There's a particular estate-planning option that we should discuss here, because it affects many family farms.

Despite increasing awareness of this estate tax planning option, Section 2032A is generally misunderstood. It is widely perceived as easy, uncomplicated, and the primary method of solving farm estate planning problems.

> **Special Land Use Valuation, Internal Revenue Code Section 2032A:** This is an alternative land valuation method used to calculate federal estate taxes.

Unfortunately, that perception often is based on a lack of accurate information concerning the complexities of 2032A and of its true advantages and disadvantages. Our purpose here is to examine Section 2032A and its potential role when being used to reduce estate taxes.

Essentially, the purpose of Section 2032A is to allow farmland to be valued as farmland. Quite simply, Section 2032A attempts to establish a productive value that is less than the farmland's fair market value (what it would bring if sold for its highest and best use). Valuing farmland at a lower price can save a significant amount of estate tax. Indeed, in select situations it can mean the difference between keeping a farm in the family or selling it to raise the cash to pay estate taxes.

For an estate to take advantage of Section 2032A, Special Land Use Valuation, it must meet the following conditions to qualify:

- The farm estate must be made up of "real property" used in farming that has a fair market value of at least 25% of the total value of the adjusted estate.
- The farm assets, both real and personal, must make up at least 50% of the estate.
- The farm real property must have been owned by the deceased or a family member for five of the previous eight years.
- The real property qualifying for special land use must pass to a qualifying heir (usually a family member).
- For five of the preceding eight years, the qualifying real property must have been farmed or there must have been material participation by the deceased or a member of the family.
- The executor (personal representative) must file an election for 2032A, with an agreement, signed by each person having an interest in the property, consenting to the liability for any estate tax recapture that may occur later.

What's material participation? There's no definitive test, but physical work, participation in a substantial number of management decisions, and employment on a full-time basis are evidence of material participation.

You might be wondering, "What does all that mean?" Let's look at an example.

Assume that your estate has a fair market value of $1 million—farmland valued at $250,000, equipment valued at $250,000, a house valued at $100,000, and personal property and investments valued at $400,000. Because your farmland constitutes 25% of your estate value and your equipment constitutes another 25%, your estate would meet both the 25% farm real estate and 50% farm assets rules.

But your estate would qualify only under the following three conditions:

- You owned the farmland and farmed it for five of the past eight years or you have cash rented it to a qualified family member.
- You transfer the farm to one or more of your children or other qualified family members, who continue to farm the land or materially participate in the farming operation.
- One of your heirs continues to farm or be materially engaged in the farming operation for another ten years.

Otherwise, any estate tax reduction is subject to recapture. If the qualifying farmland is taken out of production or sold to a non-family member during that ten-year period, there is a recapture of the estate tax reduction. In essence, the special land use valuation creates a tax lien against your farm in favor of the U.S. Treasury. A negative aspect of this election is that it makes the qualifying heir personally liable for paying the additional estate tax.

How Section 2032A Reduces Estate Taxes

Section 2032A establishes an alternative value based on a formula. The value of the land is calculated as follows:

The excess of the annual gross cash rental for comparable farmland in the same vicinity minus annual state and local real estate taxes (both determined on a five-year average) divided by the average annual effective interest rate of all new Federal Land Bank loans.

If cash rents in your area have averaged $250 for five years, real estate taxes have averaged $20 an acre, and the new average Federal Land Bank loan rate is 6.12%, we would calculate the value as follows:

$$(\$250 - \$20) / .0612 = \$3{,}758 \text{ per acre}$$

This is the alternate value of your qualified land to calculate your estate taxes. The difference between using fair market value and 2032A special use value in calculating federal estate taxes becomes the amount of the lien that can be recaptured. If the Section 2032A special land election is made, the land does not receive a total step-up in income tax basis. The basis is adjusted only to the alternate special use value. This could have adverse income tax consequences later for your heir(s) if the land is sold within the ten-year period. If a recapture occurs, there is a step-up in basis to the extent of the recapture.

A significant planning consideration arises if you intend to have one or more of your children farm, yet you also have other, non-farming children. You could have a situation where the farming children bear a greater share of the estate tax burden if a recapture occurs, because they are personally liable for taxes due at that time. Too often, the result is that the farming heirs have no choice but to have their non-active siblings as their partners for the ten-year period so they share the personal estate tax liability. Avoiding unfair treatment of your heirs requires careful consideration.

Another unexpected burden may arise when, after the 2032A election has been successfully made, the farming heirs discover that the tax lien makes lenders uncomfortable with the new debt-to-equity margins. For the farming heir, restricted borrowing capabilities could very well make the difference between the farm succeeding and failing.

As farmers age and move into retirement, they often divest themselves of farm implements and equipment by gifting or selling to their farming children and not replacing these assets. This can create a potential imbalance when planning to use the 2032A valuation.

Over time, a retired farmer can reduce farm assets and increase non-farm assets such as investments and savings. When farm assets are less than 50% of the adjusted gross estate, the estate no longer qualifies for the 2032A special land use election. Investing in more land seems attractive, but to qualify for Section 2032A the farmland must have been owned and farmed for five of the previous eight years. The strategy of buying more land also has a reduced impact in larger estates, because the valuation reduction cannot exceed $1,000,000.

So, if you believe that electing 2032A is simple and that it's the universal solution, you're mistaken. Electing 2032A involves complicated issues and many components to monitor over time to ensure that use of

2032A is consistent with your other estate plans. Here are several questions and issues to consider:

- Does your estate qualify for the 2032A election?
- Assuming that you live another ten or twenty years, will your estate still qualify for the 2032A election?
- Could your heirs successfully operate under 2032A if it were elected today? What about in ten or twenty years?
- Would restricted borrowing capabilities caused by the 2032A election significantly impact your farming heirs?
- Who would bear the burden of the potential 2032A recapture?
- Will your farming heirs want to be partners with your non-farming heirs?
- Have you discussed these issues with your personal representative (executor), farming heirs, and estate-planning attorney?
- Is the limited adjustment in income tax basis with the 2032A election a critical concern for your heirs?
- What other potential solutions should you explore?
- Is 2032A the best alternative for your estate and your heirs?

IRC Section 2032A has its place in farm estate planning. It can provide a viable alternative to other estate tax planning techniques. But it's important to understand that:

- There are potential pitfalls as well as benefits to using this election.
- There is no guarantee that a farm estate will qualify at the time the election is needed.

Many estate planners suggest that relying on the use of 2032A as the principal method to reduce estate taxes is not estate planning but rather the absence of it.

Installment Method to Pay Estate Taxes

Federal estate tax is payable in full within nine months of the date of death! But there's a break if the estate includes an interest in a closely held business. The Internal Revenue Code provides a limited alternative that allows an executor to pay a portion of the federal estate taxes over a period of years, under certain conditions.

Under Code Section 6166, the executor may elect to pay the federal estate tax allocated to the decedent's interest in a closely held business over a period of up to 14 years.

The tax itself is paid in 10 equal annual installments, beginning on the 5th anniversary of the due date for filing the federal estate tax return. The first payment of the tax coincides with the last payment of interest only, resulting in a period of 14 years. During the first five years, interest is paid on the unpaid tax. After that, annual installments of principal and interest are paid over as many as ten additional years.

IRC section 6166 allows the installment payment of the estate taxes if the value of the business exceeds 35% of the gross estate. Because the government would loan the taxpayer money at only 2.0% interest on the first $474,000 of tax due, with interest-only payments for the first five years, and interest equal to only 45% of market rates on the rest, with a payout over the next ten years, this is a very attractive option.

Qualifying for Section 6166

How does an estate qualify for IRC Section 6166? There must be reasonable cause to qualify for the extension. Unfortunately, the IRS does not clearly define reasonable cause. It examines each situation on its own merits.

However, there are guidelines used to establish reasonable cause, including the following:

- The estate is unable to raise cash or sell off assets to pay the estate tax when due.
- The estate is made up of a large percentage of accounts receivable.

- The estate has insufficient liquidity to pay the estate tax and is restricted from otherwise borrowing against the estate assets and/or accelerating the collection of the accounts receivable.

Additionally, the gross estate must consist of an interest classified as a closely held business with a value in excess of 35% of the adjusted gross estate.

To qualify as a closely held business interest, the interest can be in a sole proprietorship, a partnership, or a corporation. Partnership interests qualify if at least 20% of the total capital interest in the partnership is included in the gross estate or if the partnership has fewer than sixteen partners. Corporate stock qualifies if at least 20% of the voting stock is included in determining the gross estate of the decedent or if the corporation has fewer than sixteen shareholders.

All active business assets are considered for the 35% of adjusted gross estate test. Only active business assets qualify for the deferred payment of estate taxes. Active means that the decedent actively managed the business.

The bottom line is that consideration should be given to using 6166 when the estate cannot pay the estate taxes without liquidating land, securities, and other assets or when the estate is earning a rate significantly higher than the interest expense on the 6166 installment method.

Let's close this chapter with an important question: Why shouldn't an estate rely on Section 6166 as an alternative to sound estate planning? We have many answers to that question.

The IRS can require the estate to obtain a surety bond for up to double the amount of the tax that is being deferred. This provision has always been in the Internal Revenue Code. This means that the installment payment of estate taxes may no longer be a feasible strategy because business owners may not be able to afford the surety bond or may not be able to obtain one period. IRS officials acknowledge that a surety bond could be difficult or impossible to obtain.

Section 6166 is not reliable. The percentage test may fail. Section 6166 only defers the liability to a later date. This may still be devastating to a business operation. Interest plus the tax increases the total cost to be paid. The money to pay the estate tax is still needed. It's more economical and usually easier to predetermine the source. Usually, only a small portion of the federal estate taxes will be deferred from current payment. The executor may remain personally liable for unpaid taxes during the fourteen years and a federal tax lien will attach to the estate assets.

Distributions to heirs from the estate may be delayed up to fourteen years.So, it's definitely wisest to do comprehensive estate planning—as we've advised throughout this book. Do what is best for you and your family, and begin today.

10

Issues Farm Families Face

You have a basic understanding of the estate planning issues that many families face. The goal of this section is to help you evaluate some of the estate planning issues that are unique to farming families.

Keeping the Farm in One Piece

Farms are comprised of a relatively large capital investment for the average return generated.

Because most of the investment is in land, equipment, and buildings, farm estates generally lack liquidity.

Many farmers want to pass on their farm operations to their children. Estate and income taxes and the economics of agriculture make it very challenging to accomplish this. Keep in mind that the first estate tax rate over $1,000,000 is 41% and goes up to 55% at the top bracket.

The primary issues of farm estate planning are:
- Transferring the ownership of the farm and the farm business.

- Providing financial security for all family members involved.
- Balancing fairly and equitably the interests of farming heirs with non-farming heirs.
- Developing farm management capacity in heirs.
- Reducing estate and transfer taxes.
- Providing sufficient cash liquidity to pay estate taxes, income taxes, final expenses, debts, and mortgages without having to sell off assets necessary for the farm operation to continue.

Fair Isn't Always Equal

Many farm families have children who want to continue the farming operation. This same family may have children who are off the farm and who have decided not to be engaged in farming. Typically, this creates a dilemma for the parents: How do we compensate the farming heirs and the non-farming heirs fairly?

Is equal distribution of the assets at death fair to the children who stayed on the farm? The farming heirs may expect to receive more of the farm assets because of the greater effort they contributed. They also may not be able to continue farming without special consideration.

Additionally, the retirement transition for Mom and Dad probably was easier because the farming heirs were still on the farm.

How then is the issue of fair treatment best handled? What is the single best way to provide fair treatment for the children? Unfortunately, these questions are not easy to answer, but they are planning questions that need to be resolved for the farm estate plan to be truly successful.

What is the value of the farm?

In a farm family that has both farming and non-farming heirs, perception of the value of farmland is likely to be significantly different. Those heirs who work the land all year to earn a living realize that it has a "productive value." Their perception of value is usually lower than the heirs off the farm. Those heirs who work in town or live in the city may perceive only that the going "market value" to sell is a handsome sum. It is no wonder that confusion and problems arise in settling the farm affairs when farming parents leave the dividing up to the heirs.

The heirs may get along well while Mom and Dad are there to referee; but, when they are gone, things may be completely different. Often, it is the off-farm heirs and their spouses who get involved in discussions to determine what the going rate of cash rent is or at what price and terms to sell.

The long-term success of the farm and the farming heirs may well be dependent on parents establishing the rules and provisions in their estate plan as to how each of the heirs is to be compensated. This plan is likely to include options for the farming heirs to purchase the land from the non-farming heirs, how the price is to be established, and how the funding is to be provided for.

LIFETIME GIFTS VERSUS TRANSFERS AT DEATH

Each of us is allowed to make a gift of $13,000 a year without incurring a gift tax. This concept was discussed in greater detail in Chapter 3. This exclusion can be multiplied by the number of donors and recipients. More simply, two parents can give $26,000 to each child every year without incurring a gift tax.

A question farm parents consider: Is it best to give assets away during lifetime, or is it better to do it at death? Again, the specifics of each situation dictate the best solution.

An important consideration is the effect of having stepped-up cost basis for income tax purposes. The cost basis of an asset (for example, land) is the price that was paid for it when it was initially purchased plus any improvements. The cost basis is used to determine how much gain there is when an asset is sold. For example: A farm is purchased for $500 an acre and is sold for $5,000 an acre. This would create a taxable capital gain of $4,500 an acre that would have to be reported for income tax purposes. When an asset is transferred by gift, the original basis follows the asset.

For example: Let's assume the same acre of land is gifted to an heir/son and he sells it. He would receive $5,000 for it and have to report a $4,500 gain because he would have the same $500 cost basis as the donor/parent. However, if the same acre of land had been transferred by inheritance, the receiving heir would receive a stepped-up basis or $5,000 and could sell the land without gain. What is likely to happen when an heir receives an asset should be considered when deciding to make a lifetime transfer or make the transfers after death.

Special Land Use Valuation: 2032A

Many farm estates will be subject to Federal and Minnesota estate taxation. This creates an additional planning challenge when trying to develop a farm estate plan that will ensure the future of your family farm. However, the government did provide for a special estate taxation that principally includes farming operations.

Under these special provisions, a qualified farming estate can elect to use a special formula to value land when determining the amount of estate taxes that have to be paid. To qualify for this election, a farm estate must have farm assets (less debts and expenses) that are at least 50% of the gross estate. At least 25% of the adjusted value of the estate must be in qualified farmland and buildings. The farm real property must pass to qualified heir (usually a son, daughter, and/or spouse, although other lineal heirs may qualify). The farm real property must have been owned and operated in a farming capacity for at least five of the last eight years prior to the death of the person passing on the estate.

This valuation is calculated by a formula that takes into consideration average net farm rents for the area that includes the qualified farm real estate and the Federal Land Bank capital rate for new loans. This formula establishes a "productive value" of the farm real estate. This value is often substantially lower than market or appraised value of the same real property.

There are reduction limits and recapture rules to this special election. The reduction of the gross estate cannot exceed $1,000,000. Also, the qualified farming heirs must use the land in farming or related use for ten years or the benefits are recaptured. More simply, the full amount of estate taxes would be due if the land is not farmed for the ten-year period.

This valuation method is elected after death by the executor of the estate. From an estate-planning perspective, this election provides an important and additional planning option to reduce the amount of estate taxes paid by a farm estate.

Farm Estate Case Study: Robert and Betty Eichner

The Eichners, Robert and Betty, had five children and a substantial farm estate. Their two oldest sons, Wayne and Mark, are actively involved in farming. The rest of the children are grown and have moved off the farm. Robert, Wayne, and Mark are farming together, although Wayne and Mark are renting additional acreage that is necessary to support the entire operation.

Wayne and Mark want to own the family farm someday, but right now there isn't enough cash flow to make installment contract payments to their father, Robert. Also, Robert isn't ready to quit farming and is unsure whether it is a good idea to sell because of the capital gains income tax that would have to be paid.

Robert and Betty met with their attorney to discuss revising their wills and estate distribution plan. Their attorney suggested that they include provisions in their will that would allow Wayne and Mark the option to purchase the farm assets, including the real property if they choose to, when the estates were being settled. Robert and Betty decided to include that provision in their will. It spelled out the terms that were available to Wayne and Mark, and provided a method to determine the value of the farm assets and real property. Robert and Betty decided that holding onto the farmland best met their retirement needs.

When the Eichner's estate plan was completed, they felt good knowing that the guidelines had been established for their family to settle their estate peacefully. Wayne and Mark felt better knowing that there was a plan in place to help them acquire the family farm at some point in the future. The plan was communicated to all by Robert and Betty to ensure that none of the heirs would be surprised at the time the estate was settled. They made sure the non-farming children understood why the plan was set up the way it was.

(This is an example and should not be used as a basis for your farm estate plan. Each estate planning situation is unique. Consult with an attorney when planning your estate.)

Funding Alternatives

Often when provisions are included in an estate plan that allow the farming heirs to purchase the farming assets from the non-farming heirs, there is a need for capital funding. It is very important for farming heirs to develop a strategy to acquire the capital necessary to execute the plan.

Farming heirs have four basic ways to acquire capital necessary to execute the purchase of assets from non-farming heirs:

1. The most expensive and widely used method is to borrow. This could be from either a private or public lender. If this is the funding method chosen, it is very important for the farming heirs to have an established lending relationship and the ability to service the new debt obligation.

2. The second method is by establishing a special fund for this purpose. Excess earnings can be tunneled to a special account over a period of years to save the projected amount required to carry out the plan when needed. This method is very cost-effective compared to borrowing, but timing and time available can be important issues when determining if it is an appropriate strategy.

3. The third method is by purchasing cash value life insurance on either or both parents by the farming heirs. This method has proven to be very cost-effective; the life insurance industry has made great strides in recent years to become competitive with other saving and investment alternatives. The most important benefit of life insurance is that the

timing issue has been eliminated because the death benefit is immediately available to the farming heirs.

4. The fourth method—commonly referred to as the "rich uncle" method—is when there is a benevolent family member or other resource that will provide funds on an interest-free basis or outright gift. This method for most individuals is unavailable or unreliable, but it does deserve mention.

To determine which option is right for your family involves careful analysis, frank discussions, and difficult decisions.

A Word About Life Insurance

Keep in mind that if you own life insurance to pay for debts and estate taxes at your death, and you own it outright in your name, the death benefit is included in your gross estate for the estate tax calculation. In essence, you have named the government as one of the beneficiaries on your life insurance if you have a taxable estate. Somewhere between 47% and 55% will be paid to the government when your Form 706 "United States Estate Tax Return" is filed.

One alternative is the irrevocable life insurance trust. An irrevocable life insurance trust lets your beneficiaries benefit from the insurance proceeds and keeps the value of insurance out of your taxable estate, potentially saving your family thousands of dollars in estate taxes.

"There Is Plenty of Time"

The number one reason many farm families are faced with the hard choices of sorting out many of the farm estate issues after Mom and Dad are gone is procrastination. Facing up to our own mortality is one of the most difficult realizations in life.

Many individuals approach estate planning with the attitude that there is plenty of time to take care of it; but, unless you have extraordinary connections, you won't know when it's your time. Waiting until a death occurs in your family to begin estate planning is very perilous. If it was needed tomorrow, would your present farm estate distribution plan accomplish everything that you want and that your family needs?

When developing an estate plan, it is important to realize that there should be two parts of that plan. One part is to carry out as much during your life as is possible, comfortable, and consistent with your other planning criteria. The other is to carry out your plans when you run out of time.

Estate Planning and How to Choose an Attorney

Three main steps:

1. Determine what your estate planning objectives are. By now, you probably have a sense of some of the issues that are important to you. Gaining additional information is probably important to complete this step of estate planning. Many times estate or financial planners can be very helpful in this process. Professional designations are helpful in choosing a qualified planner, such as a CFP (Certified Financial Planner), CLU (Chartered Life Underwriter), or ChFC (Chartered Financial Consultant).

2. Inventory your assets and liabilities. This can be a tedious task. But, it is very important that you do a thorough job of determining what you own, how you own it, and what it is worth. A common tendency is to undervalue assets. Keep in mind that for estate tax purposes, the IRS is going to insist on the *highest and best use* valuation.

3. After you have taken care of the first two steps, it is time to meet with a qualified attorney. Without the use of an attorney, it is likely that your estate plan will be flawed. There is no substitute for an attorney's objective legal advice. Because the practice of law is often specialized, when meeting with an attorney, ask if estate planning is his/her area of

expertise. Attorneys are your best resource and he or she will recommend an attorney who is qualified, if estate planning is not his or her area of practice.

Additional Questions to Ask

- What are the hourly rates charged and what can I expect to pay?
- Will you be working with them personally or will the work be passed on to another firm member?
- Can and will they work with, and coordinate with, other advisors?

Remember, a well-thought-out plan is only as good as the legal work designed to carry it out. Use a qualified attorney you feel comfortable with and are confident will do his/her best to help you carry out your farm estate distribution plan.

Move Forward with Your Farm Estate Plan

Establish some goals and deadlines to have your plan reviewed, revised, and/or completed. Your farming children's long-term success may well depend on it.

The purpose of this book is to be a general guide designed to cover a broad range of essential estate planning topics. It is in a format that demystifies the topics and brings significant awareness of a wide range of issues that must be dealt with for effective, comprehensive estate planning.

Accomplishing an effective, well-thought-out estate plan is a challenging process. The techniques examined in *Keeping the Farm In Your Family* will help in making correct choices when beginning or continuing an estate planning journey.

You have worked hard for your family farm. Shouldn't it stay in *your* family?

Ten Common Mistakes in Estate Planning

1. **No will or estate plan or obsolete will**

    When you don't have a will, your estate may be distributed according to the state's intestate succession rules. In essence, the state has a will for you. However, it may not distribute your property according to your wishes. The same fate could befall your estate if your will is an obsolete will or executed in another state.

2. **Lack of specific directions in your will**

    Many wills are drafted without provisions to deal with a family business, specific assets, or other issues that require more detailed attention in the will. This happens when standard distribution language is used in the will instead of detailed instructions for the proper distribution of complex assets.

3. **Jointly owned property when there are specific will instructions**

    If your will has specific instructions for specific property and you own that property in joint tenancy, the property will pass to the joint tenant(s) instead of being distributed according to the wishes you express in your will. Property in joint tenancy always passes to the other joint tenant(s).

4. **Jointly owned property with a-b trust provisions in your will**

    If you use an estate tax reduction strategy in planning your estate and your will contains provisions for an a-b trust (credit shelter trust), joint tenant property will prevent your will from working properly. Property held in joint tenancy is not distributed by your will: It always passes to the other joint tenant(s).

5. **A-b trust provisions with large retirement plans**

    Under the employee retirement income security act of 1974 (erisa), your spouse is the required beneficiary of your retirement plan—IRAs, 401(k)s, TSAs (tax-sheltered annuities), pensions, and profit-sharing plans. Unless you follow specific steps, this beneficiary requirement overrides your will. For your retirement plans to qualify for your personal estate exemption, pay special attention to naming your beneficiary.

6. **Property owned in two or more states**

    If you own property in a state other than your state of domicile, you will have a second probate in that state to settle and transfer that real estate. One option to avoid any ancillary probate proceedings in another state is to put any real estate you own in that state into a trust.

7. **Improperly owned life insurance to pay estate taxes**

    If you've purchased life insurance to pay estate taxes and you own the policy on your own life, the death benefit is included in your gross estate to determine the amount of estate tax your estate will have to pay. One option that removes the death benefits from your taxable estate is a properly executed irrevocable life insurance trust.

8. **Undervaluing or not valuing assets when doing estate planning**

    People quite often undervalue their assets. Consequently, when the estate is being settled, estate taxes may be greater than expected. This could have very adverse results if estate assets that should be preserved must be liquidated. When valuing assets, be realistic about their current value and expected growth. Don't forget to include any expected inheritance. This will help you plan appropriately for estate tax liability.

9. **No provisions made for medical emergencies**

    Research has determined that one out of every two people will need some long-term care, and 40% of the people in this group are between the ages of 18 and 64. Causes for temporary disability or nursing home care are injuries, failing health, and illness. Estate planning should include a durable power of attorney or trust, a health-care power of attorney, and a living will.

10. Procrastination

An overdeveloped sense of longevity. Believing that there's plenty of time to do estate planning. Fear of mortality. These are three reasons why people often leave their estates in a mess to be sorted out by their heirs. Death and taxes—both are a certainty!

11

Buy-Sell Agreements

What is a Buy-Sell Agreement?

A buy-sell agreement, sometimes called a buyout agreement or a business will, is a legally binding contract between co-owners of a business that governs what happens if a co-owner dies, is otherwise forced, or chooses to leave the business. The agreement obligates one party to sell and another party to buy a particular ownership interest in a business upon certain triggering events. Individuals who are not current owners, e.g., an employee, an outsider, or a family member, may also be party to the agreement.

The buy-sell agreement specifies:

- Who can buy a departing co-owner's or shareholder's interest in the business

- What price will be paid for a co-owner's or shareholder's interest
- Which events will trigger a buyout

To ensure the buy-sell agreement is well-funded and guarantee money will be available when the buy-sell event is triggered, business succession specialists and financial planners recommend an insured buy-sell agreement.

Who Needs a Buy-Sell Agreement?

Many business owners get mired in the operations of keeping their businesses afloat today and put off planning for the future to some more convenient but distant date. And because most of us have difficulty accepting our own mortality, it's easy to justify inaction because we have virtually forever to get our ducks in a row. Right? What could possibly happen? Well, in short, many different contingencies can impact the success of your business.

Although unplanned contingencies can include loss of a professional license, insolvency, bankruptcy, or divorce, most buy-sell agreements are triggered by:

- Death
- Disability
- Retirement

Morbid as it is to consider, the probability of death of at least one of two business owners at an age prior to sixty-five is surprisingly high. The chances out of a hundred that at least one of two business owners in relatively good health (qualifies for standard insurance rates) will die prior to age sixty-five are listed on the following page.

BUY-SELL AGREEMENTS

Owner Ages	Chances out of 100
30/30	36.4
35/35	35.0
40/40	33.0
45/45	29.9
50/50	24.7
55/55	44.5
30/35	43.4
35/40	42.0
40/45	39.7
45/50	36.2
50/55	37.5

Courtesy of Number Cruncher Software

The loss of an owner creates a void in both management and operation, presenting a serious challenge to business continuation. When you plan for these contingencies, you will have a blueprint of how to proceed if that event transpires.

Every co-owned business needs a buy-sell agreement when the business is formed or as soon after that as possible. These agreements may be used by any type of business entity, sole-proprietor, corporation, partnership, or limited liability company (LLC).

A buy-sell agreement protects business owners when a co-owner leaves the company (and protects the owner who's leaving). If a co-owner wants out of the business, wants to retire, wants to sell his shares to someone else, goes through a divorce, or passes away, a buy-sell agreement acts similarly to a prenuptial agreement to protect everyone's interests, setting the price and terms for a buyout. Every day that value is added to a business without a plan for future transition, it increases the owners' financial risk.

Without a buy-sell plan, you expose your business to the following risks:
- Surviving partners or stockholders may be forced to sell or dissolve the business.
- The business may have to determine how it will compensate the decedent's family for the decedent's ownership interest.
- The decedent's estate cannot be guaranteed it will receive a fair price for the business interest.
- The business will suffer the uncertainty that accompanies losing the services of a significant member of the company.

Properly structured and funded, a buy-sell agreement can protect the wealth of business owners and their families by:
- Providing liquidity and a buyer for a departing owner's share of the business
- Avoiding conflicts between business partners
- Preventing unwanted people or family members from becoming shareholders
- Fixing the value of the business entity for estate purposes
- Not leaving an estate open to the long, expensive, and often losing process of trying to prove a lower value against a higher assessment by the Internal Revenue Service (IRS).

What is Involved in a Business Valuation?

Determining the value of your business today and at the time of a triggering event can be challenging. To assure that the business interest is transferred at a fair price, as well as help establish a value for estate tax purposes, business valuation is a critical element of the buy-sell agreement. If a business valuation is not properly completed, the IRS may challenge the value determined by the owners. Additional taxes could be levied at both the personal and business levels if courts rule in favor of the IRS.

Your business valuation should be completed by a professional appraiser. For most businesses, calculating assets minus liabilities is not comprehensive enough. Additional considerations include:

- Average annual earnings
- Number of years in business
- Nature of the business
- Ownership control
- The earning power of a business
- Goodwill
- Earnings of comparable companies

Once the business valuation has been completed, it's important to review it periodically and adjust valuation as needed to reflect any significant change in business operation or value.

Like any other property included in a decedent's estate, the value of a business interest is assessed at fair-market value for estate tax purposes. Fair-market value is the hypothetical price that a willing buyer would pay and that a willing seller would accept when neither is under any compulsion to buy or sell and both have reasonable knowledge of the relevant facts.

Types of Buy-Sell Agreements

Entity Plan

An Entity Plan buy-sell agreement transpires between a business entity and its owners. Say a business is equally owned by Pat and Terry. Both Pat and Terry enter into an agreement with the business for the purchase and sale of their respective interests. The agreement obligates both Pat and Terry and their estates to sell, and the business to buy, upon the death, disability or retirement of either of them.

If Pat dies, becomes disabled, or retires, the agreement transfers Pat's ownership interest in exchange for cash. Once the cash is received by Pat's estate, the business interest is transferred to the business.

Life Insurance-Funded Entity Plan.

The business invests in a life insurance policy on both Pat and Terry. This way, the business is the owner, premium payer and beneficiary of each policy. When Pat or Terry dies, the business receives the funds and uses the money to buy Pat's or Terry's interest in the company.

Advantages	Disadvantages
• One life insurance policy per owner funds the plan. • The business, rather than the owners, bears the burden of premium payment. • When an owner dies, the value of the interest held by the surviving owners is increased by the purchase price. • Policy proceeds are generally received income tax-free by the business.	• Policy proceeds and accumulated values are also available to creditors of the business. • Life insurance payout will "balloon" the value of the business before the buyout. • For C Corporations, policy proceeds could be exposed to the alternative minimum tax.

Cross Purchase Plan

A cross purchase buy-sell agreement provides for the planned disposition of owner interests at the death, disability or retirement of a co-owner. Pat and Terry, who own equal interest in the business, both enter into an agreement to provide for the purchase and sale of their individual interests. This binding agreement obligates both Pat and Terry or their representatives to either buy or sell upon the death, disability or retirement of either one.

Life Insurance-Funded Cross Purchase Plan.

Each co-owner is required to purchase a proportional share of the deceased co-owner's interest. The number of policies required depends on the number of co-owners. Each co-owner buys a policy on all of the other co-owners. For instance, if there are three co-owners, each co-owner buys policies on the other two co-owners. So, three co-owners would require a total of six policies. Each policy amount reflects the respective owner's interest. Policy owners are also the named beneficiaries. When a co-owner dies, surviving co-owners receive proceeds from the policy income tax-free. In exchange for a payment equal to the value of the deceased co-owner's interest, the deceased co-owner's estate transfers the business interest to the remaining co-owners. The cross purchase agreement is funded by the proceeds from the policy.

Advantages	Disadvantages
• A well-calculated plan will provide the deceased owner's estate with a fair price for the business interest. • The co-owners control the agreement. • The established purchase price becomes part of the remaining owner's cost basis. • If this business is sold later, this should result in a lower taxable gain. • Unequal ownership interests may be retained or changed.	• Gets complicated quickly with more than a few co-owners. • Co-owners purchase policies with their personal funds. • If the business valuation increases, original policies may prove inadequate.

Trusteed Cross Purchase Plan

A trusteed cross purchase plan is different in that it involves an escrow agent in the buy-sell agreement to carry out the planned disposition of co-ownership interests in case of a death, disability, or retirement.

Life Insurance-Funded Trusteed Cross Purchase Plan

An impartial trustee or escrow agent acts as custodian for the insurance policies. Most often, the trustee is the owner and the beneficiary of the policy. The trustee may collect premium payments from the business co-owners, as stipulated in the buy-sell agreement. When a co-owner dies, the trustee distributes the proceeds to the deceased co-owner's estate in exchange for the value of the business interest.

Advantages	Disadvantages
• If a funding problem arises with a co-owner, the trustee can advise other co-owners of the situation. • The trustee receives policy proceeds tax-free. • The trustee distributes policy proceeds to the deceased co-owner's estate in exchange for the ownership interest.	• Potential transfer-for-value issues could result in higher income taxes at the next co-owner's death.

Key Person or One-Way Plan

A key employee and co-owner enter into a buy-sell agreement. At the co-owner's death, disability, or retirement, they key employee buys the co-owner's business interest.

Life Insurance-Funded Key Person Plan

The key employee buys a life insurance policy on the co-owner and pays the premium out of personal funds. The key employee owns and is the beneficiary of the policy. When the co-owner dies, the key employee buys the deceased co-owner's business interest from the estate.

Advantages	Disadvantages
• The key employee has a vested interest in the prosperity of the business. • Provides high level of control with relatively little administration. • Co-owner's estate should receive a fair price for the business interest.	• Key employee must pay premiums. • Premiums are not tax deductible. • Premiums can be high if co-owner is unhealthy or elderly.

Buy-Sell Funding Alternatives and Impacts

Generally, when a buy-sell agreement is executed, one party transfers cash in exchange for all or a portion of the business. Ideally, the method used to fund a buy-sell agreement should:

- Have a relatively low cost
- Be easy to understand
- Be easy to establish, maintain, and execute
- Not diminish working capital or creditworthiness

Although most buy-sell agreements are funded with life insurance, alternatives exist to create the liquidity necessary to purchase a business interest. The most common alternatives and their impacts are summarized below.

Cash

The first alternative is cash. The considerations for funding a buy-sell agreement with cash include the following:

Advantages	Disadvantages
• Simple. • Requires no immediate outlay.	• Can't predict when or how much cash will be needed. • Must always keep adequate cash reserves on hand. • Survivor's liquid assets could be depleted. • Could drain corporate surplus, operating capital, or current income. • Have to use after-tax dollars. • Money held in reserve is not likely to earn a return. • Accumulated cash for a buyout could trigger an accumulated earnings tax. • May have to increase survivor salaries to enable payments of both principal and interest.

Savings Account

Under the savings account plan, annual deposits are made to a savings account. Deposits are intended to accumulate a sum equal to the purchase price of the individual business interests. For a cross purchase plan, the amount of the savings should equal the value of the other ownership interests. For an entity plan, the business makes deposits in the savings account equal to the individual ownership interests.

Advantages	Disadvantages
• Can be used if one of the owners is uninsurable.	• A low-risk, low-yield investment vehicle must be used to ensure the safety of the principal. • Annual earnings on cross purchase plans and entity plans may be subject to income tax to be paid by the account owner. • Under an entity plan, the savings account is considered a business asset and is subject to the claims of creditors. • Because death or long-term disability is almost always "premature," savings funds are often inadequate. • May lack the requisite time or discipline to build up and maintain sufficient savings. • Decedent's family must rely on the financial ability and moral obligation of surviving co-owners.

Loan

Like the cash alternative addressed earlier, a loan is easy to understand but the number of disadvantages don't make this a viable alternative in most instances.

Advantages	Disadvantages
• Simple. • Requires no outlay until death or disability occurs.	• Could be tough to get a loan when a critical contributor is out of the picture. • Cost to borrow is substantial. • Loan terms and rates may not mesh with what the borrower can afford. • Demands of repaying the loan may compromise operation and creditworthiness. • Loan interest on an installment cross purchase may not qualify as investment interest. If that's the case, the whole deduction could be lost.

Installment Payout

Although an installment payout may seem to be a reasonable means to purchase a deceased co-owner's interest, the risks are considerable.

Advantages	Disadvantages
• Simple. • Nothing needed until the seller's death. • Relatively small outlay is required each year. • Heirs receive interest on unpaid balance. • Gain on the sale may be spread over a number of years.	• Does not provide large sums of cash typically needed for estate settlement costs and debt. • Spreads out the obligation but does not provide cash for a buyout. • May not provide decedent's family with desired financial security. • Could adversely affect creditworthiness. • Could pose IRS problems. • If IRS deems rate of interest on unpaid balance inadequate, IRS will "impute" higher fair-market interest rate. • Corporate assets used to secure unpaid balance may impede ability to borrow money from other sources. • If corporate assets are insufficient, seller may also require personal guarantees from co-owners. • Installment obligation longer than fifteen years may not be possible.

Private Annuity

In lieu of a lump sum payout, a withdrawing or retiring co-owner can agree to receive a private annuity in exchange for stock.

Advantages	Disadvantages
• Business or buying co-owners pay a fixed amount annually to the selling co-owner for the rest of his or her life (+ or–spouse). • Allows seller to divest all stock. • Successor co-owners (family members/loyal employees) benefit.	• Cannot be "secured" without the seller losing the two main advantages of a private annuity: the removal of the current value of the stocks and future appreciation from the seller's estate, as well as the spreading out of gain over the lifetime of the seller. • Requires seller's complete trust in successor's management. • Seller may not receive payments if corporation is unable or unwilling to pay. • Interest paid by the corporation is not deductible. • Seller might die early and lose some of the value of the sale. • If seller lives beyond normal life expectancy, total payments can exceed purchase price. • Could be serious tax consequences if payments or valuation of stock doesn't jive with the IRS.

Mortgaging the Business

Depending on when the buyout must be executed, business assets may be mortgaged.

Advantages	Disadvantages
• Business assets may be mortgaged to raise cash for the buyout.	• Debt will affect the company's creditworthiness. • Servicing the debt places additional strain on the company's finances.

Life Insurance

At the death of a co-owner the estate receives a payment in the amount of the ownership interest. In exchange for the cash payment, the decedent's estate transfers the ownership interest to the business or remaining co-owners.

Advantages	Disadvantages
• Guarantees that death will provide the cash to manage business continuation needs. • Decedent's spouse/heirs get fair-market value of the business interest. • Spouse/heirs can depart business before it starts losing value. • Because cumulative premiums are typically a small percentage of the total death benefit, the amount paid for premiums is generally far below the purchase price. • Provides proof to creditors that co-owners are financially responsible. • Premiums can be budgeted so buyout doesn't hinder cash flow. • Provides peace of mind. • If lifetime buyout occurs, purchase price can be offset by policy cash values. • Policy loans and amounts received from withdrawals or partial surrenders are generally income tax-free up to the policy owner's basis. • Variety of products exist to match client's budget, risk tolerance, and time horizon.	• Premium is not tax deductible. • Requires cash outlay up front. • Premium amount can't be invested elsewhere. • If co-owners are sick or elderly the cost can be high. • Certain partial surrenders and withdrawals during the first fifteen policy years may be taxable.

How Much Insurance is Enough?

Generally, the value of a business interest is expected to increase due to growth. Inflation can also boost value. This in turn increases the buyer's liability in purchasing a withdrawing co-owner's share in the business. For these reasons, buy-sell agreements that are fully-funded (sufficient to pay off the entire selling price) provide the greatest security.

The number of years it takes for stock to double in value can be determined by dividing the growth rate into 72 ("The Rule of 72"). If the value of the business grows at 5% per year, the business value will double every 14.4 years. A 10% growth rate doubles the business value every 7.2 years.

Present Business Value: $1 Million		
Annual Growth 5%	10 Years Hence $1,628,895	20 Years Hence $2,653,298
Annual Growth 10%	10 Years Hence $2,593,742	20 Years Hence $6,727,500

The price the buyer must pay for stock can easily become greater than the insurance available to finance the obligation. Additionally, each corporate redemption or purchase by co-owners upon the death of another co-owner increases the value of the surviving co-owners' shares. As such, insurance coverage should be increased after every redemption or purchase.

Surviving co-owners of a cross purchase plan gain a larger percent interest in the same size business. If the business is the buyer, the insurance payout prevents business assets from shrinking in proportion to the stock purchased.

Intentionally Defective Grantor Trust

The IDGT is a flexible and effective method to reduce transfer taxes, as well as a great means to implement the succession plan of a closely-held business to the next generation.

What is an Intentionally Defective Grantor Trust (IDGT)?

A defective grantor trust is an irrevocable trust for the benefit of children and grandchildren where the grantor does not retain any income interest. The trust is made defective by "flawing" so that it includes income tax but excludes gift tax and estate tax. Although the spouse may be the trustee, the grantor may not be the trustee.

How Does an IDGT Work?

The difference in treatment between income tax laws and estate tax laws enables grantors to use IDGTs for income tax purposes without the consequences of gift and estate taxes. Under grantor trust rules, the trust is structured so that the grantor is treated as the owner of the trust for income tax purposes only. The grantor trust rules do not apply for gift tax and estate tax purposes. So, let's say you're the grantor. Provided that you do not retain any interest, power, or right that would cause inclusion, you would be taxed on the trust income but trust assets would not be included in your estate.

As the grantor, you sell assets to the trust in exchange for a promissory note with interest. The note can be either self-liquidated over the note's term, or can require payments of interest only, with a balloon payment of principal at maturity. Assets may include:

- Stock in a closely-held or family business
- Real estate
- Marketable securities
- Limited partnership interests

The trust should have "substance" if initially funded with assets equal to at least 10% of the value of the property to be purchased. The IDGT works best when:

- Sold assets are subject to discounts in determining their fair market value

- Sold assets are expected to appreciate at a higher rate than the interest rate payable on the note.

The term of the promissory note will be a specific number of years, such as ten or fifteen. If the IDGT purchases life insurance on the grantor's life, the note should be interest-only with a balloon payment. If possible, the note should be renewed and kept outstanding until death. Then, when the grantor dies, only the fair-market value of the note is included in the grantor's estate. The value of the note will be lower than the remaining principal relative to the payout of the note, interest rate, absence of security, default provisions, covenants, and other specified terms.

What Are the Benefits of an IDGT?

A sale to the IDGT is not recognized for income tax purposes because the grantor and the trust are viewed as the same entity by the IRS. Because the IRS does not recognize a gain or loss in transactions between the grantor and the trust, the sale of assets to the trust has no income tax consequences.

This unique taxing structure provides planning opportunities to achieve gift- and estate-tax benefits for the grantor. The most significant of these benefits is to "freeze" the value of a closely held business interest or real estate owned by the grantor. Any increase in the value of the sold assets will not be taxed in the grantor's estate and will result in greater benefit to the beneficiaries.

Defective trust planning substantially shrinks the estate while it transfers asset appreciation from parents to children and grandchildren. Even though grantors are taxed on income they don't receive, they effectively reduce their estate by the amount of the tax. By paying the tax on trust income, the grantor provides a tax-free gift to the beneficiaries. The trust also benefits from using pre-tax dollars to pay the promissory note.

While the defective grantor trust is a clever concept, it does have inherent risks. These risks can be minimized with proper planning and drafting of legal documents by attorneys qualified in this area.

12

Medical Assistance—
What You Need to Know

A 2005 study showed that the average cost for a private nursing home room increased 5.7% from 2004 to 2005. That means in most states, it exceeds $200 a day. Costs of in-home care weren't spared either, increasing 5.5%.

MetLife's Mature Market Institute

The total amount spent on long-term care services in the United States (in 2005) was $206.6 billion. This does not include care provided by family or friends on an unpaid basis (often called "informal care"). It only includes the costs of care from a paid provider.

The average costs in the United States (in 2008) were:
- $187/day for a semi-private room in a nursing home
- $209/day for a private room in a nursing home

- $3,008/month for care in an Assisted Living Facility (for a one-bedroom unit)
- $29/hour for a Home Health Aide
- $18/hour for Homemaker services
- $59/day for care in an Adult Day Health Care Center

—U.S. Department of Health and Human Services

Providing for your own health care as you age has become a complex process. Nursing home costs continue to rise and medical assistance regulations continue to get more complicated. This is why detailed and knowledgeable planning is critically important to everyone. To create the most advantageous Medical Assistance plan, the sooner you start, the better your chance of making the most of both your assets and Medical Assistance/Medicaid benefits.

The following is an overview of basic Medical Assistance planning. If you do any Medical Assistance planning or any other aspect of estate planning, you should always seek the advice of professionals (attorneys, accountants, financial planners, and trust officers).

Medicare

(Adapted from the Centers for Medicare and Medicaid Services and AARP.org) Medicare is a health insurance program for people in the following categories:

- age 65 or older,
- under age 65 with certain disabilities, or
- of all ages with End-Stage Renal Disease (permanent kidney failure requiring dialysis or a kidney transplant).

Medicare offers hospital insurance, medical insurance, and prescription drug coverage.

Part A: Hospital Insurance

Most people don't have to pay a monthly payment, called a premium, for Part A. This is because they or a spouse paid Medicare taxes while working. Beneficiaries who don't get premium-free Part A may be able to buy it if they (or their spouse) aren't entitled to Social Security, because they:

- Didn't work or didn't pay enough Medicare taxes while working,
- Are age 65 or older, or
- Are disabled but no longer get free Part A because they returned to work.

After you have paid your hospital deductible ($1,068 in 2009), the Original Medicare plan pays all your hospital costs for up to 60 days in a benefit period. A benefit period begins the day you go to the hospital and ends when you have been out of the hospital for 60 days in a row. If you go into the hospital again after 60 days have passed, you begin a new benefit period.

If you stay in the hospital more than 60 days, you pay $267 (in 2009) a day for days 61 through 90. If you stay longer than 90 days in a benefit period, the cost for each day is $534 (in 2009) for up to 60 days over your lifetime.

States may help people with limited income and resources to pay for Part A. For more information, visit www.socialsecurity.gov on the web or call the Social Security Administration at 1-800-772-1213. TTY users should call 1-800-325-0778.

Part B: Medical Insurance

Most people pay a monthly premium for Part B. Medicare Part B (Medical Insurance) helps cover doctors' services and outpatient care. It also covers some other medical services that Part A doesn't cover, such as some of the services of physical and occupational therapists, and some home health care. Part B helps pay for these covered services and supplies when they are medically necessary.

The Medicare Part B premium carries a monthly charge ($78.20 per month in 2005). In some cases, this amount may be higher if beneficiaries didn't sign up for Part B when they first became eligible.

A word of warning: If the beneficiary didn't take Part B when first eligible, the cost of Part B will go up 10% for each full twelve-month period that the beneficiary could have had Part B but didn't sign up for it, except in special cases. They will have to pay this penalty as long as they have Part B.

Beneficiaries also pay a Part B deductible each year before Medicare starts to pay its share. The Part B deductible for 2009 is $135.00. Beneficiaries may be able to get help from their state to pay this premium and deductible. Medicare deductible and premium rates may change every year in January.

After you pay your yearly Part B deductible, Medicare generally pays 80% of doctor and other medical services. It pays 50% of mental health services and 100% of some preventive services.

Medigap plans (next page) cover all or part of your share of the services mentioned above—20% of the Medicare-approved amount for doctor services and 50% for mental health services. The "Medicare approved amount" is the amount that Medicare decides is a reasonable payment for a medical service.

Prescription Drug Coverage

On January 1, 2006, new Medicare prescription drug coverage became available to everyone with Medicare. Most people will pay a monthly premium for this coverage. This coverage may help lower prescription drug costs and help protect against higher costs in the future.

Medicare Prescription Drug Coverage is insurance. Private companies provide the coverage. Beneficiaries choose the drug plan and pay a monthly premium. Beneficiaries who decide not to enroll in a drug plan when first eligible may pay a penalty if they choose to join later.

Medigap

Medicare has several gaps and doesn't pay for all of the health-care services you may need. If you are in the Original Medicare Plan, you may want to buy Medicare supplemental insurance, also called Medigap.

Medigap insurance is sold by private insurance companies. By law, companies can offer only twelve standard Medigap insurance plans named A through L. Each plan has a different set of benefits. Study all the Medigap plans before deciding which is best for you.

All plans with the same letter cover the same benefits. For instance, all Plan C policies have the same benefits no matter which company sells the plan. However, the premiums can vary.

In addition to the standard A through L Medigap policies, Medicare SELECT is a type of Medigap policy that can cost less than standard Medigap plans. However, you can only go to certain doctors and hospitals for your care. Check with your state insurance department to find out whether or not Medicare SELECT policies are available in your State. NONE of the standard Medigap plans cover:

> **Medigap Insurance:** Supplemental insurance to help pay for some of your costs in the Original Medicare program and for some care it doesn't cover.

- Long-term care to help you bathe, dress, eat, or use the bathroom
- Vision or dental care
- Hearing aids
- Private-duty nursing
- Prescription drugs

If you live in Massachusetts, Minnesota, or Wisconsin, you have different standard Medigap plans. Check with your state insurance department or the *Guide to Health Insurance for People with Medicare: Choosing a Medigap Policy.*

Basic Benefits

All twelve Medigap plans pay your costs for days 61 through 150. In addition, once you use your 150 days of Medicare hospital benefits, all Medigap plans cover the cost of 365 more hospital days in your lifetime. Plan A is the most basic plan.

Plans B through L offer everything in Plan A and provide additional coverage at an additional cost. Plans K and L offer similar services as Plans A through J, but cost-sharing for the basic benefits is at different levels.

- The Original Medicare plan doesn't cover the first three pints of blood you need each year. Plans A through J pay for these first three pints.

- Plans F and J also have a high-deductible option. You will have a lower premium with the high-deductible option, but you will have to pay more out of pocket before the policy will begin to pay benefits. Plans F and J require that you first pay your annual Medigap deductible before your costs will be covered.

- If you have Plans K or L, you will have to pay a portion of the hospital deductible ($1,068 in 2009) before your costs will be covered—unless you have already met the annual out-of-pocket maximum for the year.

Extra Benefits

- **Medicare Part A Hospital Deductible:** Medigap Plans B through J cover the hospital deductible ($1,068 in 2009) for each benefit period. This benefit usually saves you money if you have to stay in the hospital.

- **Skilled Nursing Home Costs:** The Original Medicare Plan pays all of your skilled nursing home costs for the first twenty days of each benefit period. If you are in a nursing home for more than twenty days, you pay part of each day's bill.

Medigap Plans C through J pay your share of the bill ($133.50 a day in 2009) for days 21 through 100. Neither Medicare nor any Medigap plan pays for any skilled nursing home stay longer than 100 days in a benefit period.

- **Medicare Part B Deductible:** You must pay a deductible each year for doctor and other medical services before Medicare pays. Medigap Plans C, F, and J pay this deductible. In 2009, the deductible is $135.

- **Medicare Part B Excess Charges:** When you see that a doctor doesn't "accept assignment," it means he or she does not accept Medicare's approved amount as payment in full. The doctor can charge you up to 15% more than Medicare's approved amount.

    Medigap Plans F, I, and J pay 100% of these excess charges. Medigap Plan G pays 80% of the excess charges. You might want this benefit if you don't know whether the doctors you see accept assignment, such as when you are in the hospital.

- **Foreign Travel Emergency:** Medicare does not cover any health care you receive outside of the United States. Medigap Plans C through J cover some emergency care outside the United States. After you meet the yearly $250 deductible, this benefit pays 80% of the cost of your emergency care during the first sixty days of your trip. There is a $50,000 lifetime maximum.

- **At-Home Recovery:** Medicare covers some skilled home care given by a nurse or a physical, occupational, or speech therapist. It does not pay for at-home help for the activities of daily living, such as bathing and dressing. You pay for this type of care.

Medigap Plans D, G, I, and J cover this type of at-home help if you already are receiving skilled home health care that is covered by Medicare. These plans cover at-home help for up to eight weeks after you no longer need skilled care. However, they will not pay more than $40 per visit, seven visits a week, or $1,600 each year.

- **Preventive Care:** Medigap Plans E and J offer this benefit, which is limited to $120 each year. It helps pay for preventive care not covered by Medicare. Since Medicare now covers more preventive care, make sure this benefit is helpful to you.

- **Prescription Drugs:** After Jan. 1, 2006, you cannot purchase new Medigap policies covering prescription drugs because private companies approved by Medicare offer this coverage separately. In order to get prescription drug coverage, you must enroll in a Medicare Prescription Drug Plan.

- **Plans K and L:** Important: Plans K and L offer similar coverage as plans A through J, but the cost-sharing for the benefits is at different levels and has annual limits on how much you pay for services. The out-of-pocket limits are different for Plans K and L and will increase each year for inflation. In 2009, the out-of-pocket limit was $4,620 for Plan K and $2,310 for Plan L.

- **Ongoing Coverage:** Once you buy a Medigap plan, the insurance company must keep renewing it. The company can't change what the policy covers and can't cancel it unless you don't pay the premium. The company can increase the premium and should notify you in advance of any increases.

Medicare Questions Answered

1-800-MEDICARE	1-800-633-4227 TTY 1-877-486-2048
Social Security To get a replacement Medicare card; change your address or name; get information about Part A and/or Part B eligibility, entitlement, and enrollment; apply for "extra help" with Medicare prescription drug costs; and report a death.	1-800-772-1213 TTY 1-800-325-0778
Coordination of Benefits Contractor To get information on whether Medicare or your other insurance pays first.	1-800-999-1118 TTY 1-800-318-8782
Department of Defense To get information about TRICARE. To get information about TRICARE for Life.	1-888-363-5433 1-866-773-0404 TTY 1-866-773-0405
Department of Veterans Affairs If you are a veteran or have served in the U.S. military.	1-800-827-1000 TTY 1-800-829-4833
Office of Personnel Management To get information about the Federal Employee Health Benefits Program for current and retired Federal employees.	1-888-767-6738 TTY 1-800-878-5707
Railroad Retirement Board (RRB) If you have benefits from the RRB, call them to change your address or name, enroll in Medicare, replace your Medicare card, and report a death.	Local RRB office or 1-877-772-5772
Quality Improvement Organization (QIO) To ask questions or report complaints about the quality of care for a Medicare-covered service.	Call 1-800-MEDICARE to get the telephone number for your QIO.

Medicaid

Unlike Medicare, people are not entitled to receive Medical Assistance/Medicaid later in life. It is a means-based program that has asset qualifications and other limits that must be anticipated. Medical Assistance/Medicaid is governed both by the State and Federal government, making the number of regulations affecting the program substantial. Generally speaking, the Federal government provides the money and the State governments provide the administration and disbursement functions to deliver the required aid to the public in need.

Nursing home laws, Medical Assistance/Medicaid qualifications, legalities around asset requirements, and exceptions all seem to multiply each year. There are special rules for those who live in nursing homes and for disabled children living at home. It can be daunting to try to get a realistic idea of what to expect, and how to start positioning yourself now to get the aid you may need in the future.

> **Medicaid:** A joint federal/state program that provides medical care to the needy. The federal government sets minimum standards that all state Medicaid programs must meet and then each state decides what its programs will cover.

Overview of Basic Medical Assistance Planning

Medicaid helps low-income persons of all ages pay for medical and long-term care. It may also help people who have extremely high medical bills or need to pay for nursing home care. Medicaid does not pay money to you; instead, it sends payments directly to your health-care providers. Depending on your state's rules, you may be asked to pay a small part of the cost (co-payment) for some medical services.

Medicaid must pay for some services, such as inpatient and outpatient hospital services, physician and certified nurse practitioner visits, laboratory tests and x-rays, nursing home and home health care, and certain screenings. It may also pay for services, such as prescription drugs, clinic visits, prosthetic devices, hearing aids, dental care, eye exams, glasses, transportation for medical care, and medical services not covered by Medicare. Additionally, it can help pay Medicare costs.

ELIGIBILITY

To qualify for Medicaid, you must meet the income and resource guidelines in your state. Income is money you get from Social Security, a job, pension, or other sources. Resources are things you own, such as a savings account.

Many groups of people are covered by Medicaid. Even within these groups, though, certain requirements must be met. These may include your age, whether you are pregnant, disabled, blind, or aged; your income and resources; and whether you are a U.S. citizen or a lawfully admitted immigrant.

Your child may be eligible for coverage if he or she is a U.S. citizen or a lawfully admitted immigrant, even if you are not (however, there is a five-year limit that applies to lawful permanent residents). Eligibility for children is based on the child's status, not the parent's. Also, if someone else's child lives with you, the child may be eligible even if you are not because your income and resources will not count for the child.

In general, you should apply for Medicaid if your income is low and you match one of the descriptions of the Eligibility Groups. (Even if you are not sure whether you qualify, if you or someone in your family needs health care, you should apply for Medicaid and have a qualified caseworker in your state evaluate your situation.)

A person who needs Medical Assistance starts the process by contacting his or her county human services agency. After submitting the application, he or she must meet with a financial worker, who will explain the program and may ask for more information and verification regarding the applicant's situation. If the applicant is not found eligible, he or she may apply again at any time.

For state guidelines and other Medicaid information, visit the Centers for Medicare and Medicaid Services (CMS) website at www.cms.hhs.gov. See the list of toll free numbers and call the one for your state.

When Eligibility Starts

Coverage may start retroactive to any or all of the three months prior to application, if the individual would have been eligible during the retroactive period. Coverage generally stops at the end of the month in which a person's circumstances change. Most states have additional "State-only" programs to provide medical assistance for specified persons who do not qualify for the Medicaid program. No Federal funds are provided for State-only programs.

Long-Term Care

Federal law requires states to provide nursing home coverage. All states provide community Long-Term Care (LTC) services for individuals who are Medicaid eligible and qualify for institutional care. Most states use eligibility requirements for such individuals that are more liberal than those normally used in the community.

Preventing Spousal Impoverishment

The expense of nursing home care, which ranges from $4,000 to $6,000 a month or more, can rapidly deplete the lifetime savings of elderly couples. In 1988, Congress enacted provisions to prevent what has come to be called "spousal impoverishment," which can leave the spouse who is still living at home in the community with little or no income or resources. These provisions help ensure that this situation will not occur and that community spouses are able to live out their lives with independence and dignity.

Resource Eligibility

The spousal impoverishment provisions apply when one member of a couple enters a nursing facility or other medical institution and is expected to remain there for at least thirty days. When the couple applies for Medicaid, an assessment of their resources is made. The couple's resources, regardless of ownership, are combined.

The couple's home, household goods, an automobile, and burial funds are not included in the couple's combined resources. The result is the couple's combined countable resources. This amount is then used to determine the Spousal Share, which is one half of the couple's combined resources.

To determine whether the spouse residing in a medical facility meets the state's resource standard for Medicaid, the following procedure is used: From the couple's combined countable resources, a Protected Resource Amount (PRA) is subtracted. The PRA is the greatest of:

- The Spousal Share, up to a maximum of $109,560 in 2009;
- The state spousal resource standard, which a state could set at any amount between $21,912 and $109,560 (in 2009);
- An amount transferred to the community spouse for her/his support as directed by a court order; or
- An amount designated by a state hearing officer to raise the community spouse's protected resources up to the minimum monthly maintenance needs standard.

After the PRA is subtracted from the couple's combined countable resources, the remainder is considered available to the spouse residing in the medical institution as countable resources. If the amount of countable resources is below the State's resource standard, the individual is eligible for Medicaid. Once resource eligibility is determined, any resources belonging to the community spouse are no longer considered available to the spouse in the medical facility.

Income Eligibility

The community spouse's income is not considered available to the spouse who is in the medical facility, and the two individuals are not considered a couple for income eligibility purposes. The state uses the income eligibility standard for one person rather than two, and the standard income eligibility process for Medicaid is used.

Post-Eligibility Treatment of Income

This process is followed after an individual in a nursing facility/medical institution is determined to be eligible for Medicaid. The post-eligibility process is used to determine how much the spouse in the medical facility must contribute toward his/her cost of nursing facility/institutional care. This process also determines how much of the income of the spouse who is in the medical facility is actually protected for use by the community spouse.

The process starts by determining the total income of the spouse in the medical facility. From that spouse's total income, the following items are deducted:

- A personal needs allowance of at least $30;

- A community spouse's monthly income allowance (between $1,750 and $2,739 for 2009), as long as the income is actually made available to her/him;

- A family monthly income allowance, if there are other family members living in the household; and

- An amount for medical expenses incurred by the spouse who is in the medical facility.

The community spouse's monthly income allowance is the amount of the institutionalized spouse's income that is actually made available to the community spouse. If the community spouse has income of his or her own, the amount of that income is deducted from the community spouse's monthly income allowance. Similarly, any income of family members, such as dependent children, is deducted from the family monthly income allowance.

Once the above items are deducted from the institutionalized spouse's income, any remaining income is contributed toward the cost of his or her care in the institution.

Complications

Some assets are counted, while others are exempt or unavailable. There are ways to spend down the countable assets, transfer assets, or make them unavailable. But there are also restrictions on what can be done to reach eligibility level. However, a well-conceived spend-down or asset transfer plan can save most of the family unit's resources.

Understanding Medicaid law and working with qualified attorneys and planners at the beginning of the process is the best approach. This will allow a family with excess assets to convert assets based on the Omnibus Budget Reconciliation Act (OBRA) passed in 1993. It is complicated, but necessary to plan if asset preservation is important. The social worker works for the county and will not show you how to preserve assets. In many cases, the social worker doesn't understand what can be done.

For example, a family has $250,000 of cash assets (countable assets). The social worker understands that the sick spouse cannot go on Medicaid until the family's exempt assets are at $109,560 or below ($3,000 for the sick spouse and $109,560 for the healthy spouse). So the family believes it must spend down to $109,560 of exempt assets.

A skilled professional in this area would explain to the client that the $140,440 of excess countable assets ($250,000 minus $109,560) could be transferred to the healthy spouse. The healthy spouse would then purchase a qualifying Medicaid annuity and turn the annuity into a qualifying OBRA '93 period certain income stream based on his or her age. The asset is now protected and the ill spouse is immediately eligible to qualify for medicaid.

Now let's consider various assets and how MA staff would treat them in terms of the MA regulations and under the Spousal Impoverishment Act, a law passed in 1988 to protect the spouse of an applicant for MA by allowing him or her to keep certain amounts of assets and income.

How Medical Assistance Treats Assets

Medical Assistance requires that the couple list all their assets, regardless of whose name they are in, who earned them, or how long either has owned them, including any assets that were transferred within a specified period. (This period is the last thirty-six months for transfers to any person other than the spouse or the last sixty months for any transfers to a trust.) MA staff then categorize all those assets as exempt or countable or unavailable. The stay-at-home spouse is then allowed to keep half of the non-exempt assets. This share is known as the Community Spouse Resource Allowance (CSRA).

The applicant and his or her spouse must then either spend down or make unavailable any assets in excess of the exemption and the CSRA. Once they've passed the asset test, Medical Assistance then reviews their income and determines the applicant's share of the cost.

Assets That Don't Count

An MA applicant is allowed to keep the following assets, which are not counted toward the maximum:

- Household goods
- Clothing
- Jewelry
- Burial space items such as a grave marker, grave site, crypt, mausoleum, vault, casket, urn, or other repository
- A burial account
- Interest in a burial account and burial space items
- A motor vehicle (with some exceptions)

Countable Assets

The following assets are counted toward the maximum amount allowed by:

- Cash on hand
- Bank accounts
- Stocks, bonds, savings certificates
- Contract for deeds for which you hold the title

- Non-homestead property
- Extra motor vehicles
- Boats

About Income

The institutionalized spouse who receives Medical Assistance is allowed to retain a personal needs allowance and a monthly premium to pay for medical insurance. (The personal needs allowance will increase each January 1 by the same percentage as the Social Security cost of living adjustment.) All other income is paid to the nursing home except what is needed to pay health insurance premiums, deductibles, or co-payments.

There are some exceptions.

Under the Spousal Impoverishment Act, the stay-at-home spouse is allowed a Minimum Monthly Maintenance Needs Allowance that is adjusted annually for inflation.

If the stay-at-home spouse has less than the needs allowance, Medical Assistance will reduce the confined spouse's share of the cost to bring the stay-at-home spouse's income up to the minimum. If the stay-at-home spouse is not receiving any income from the institutionalized spouse, then the stay-at-home spouse may have unlimited income.

Personal Residence

The personal residence is an exempt asset if the MA applicant intends to return home and can reasonably be expected to do so. If the applicant is single and the personal residence is part of his or her probate estate, the personal residence may be subject to asset recovery by the state. If the applicant is married, his or her home is subject to asset recovery when the spouse dies.

The personal residence is also exempt if it's the residence of any of the following people:

- Spouse
- Child under age 21 or blind or disabled
- Brother or sister who owns equity in the home and lived with the applicant for at least one year just before he or she entered the nursing home
- Child or grandchild who lived with the applicant for at least two years just before he or she entered the nursing home and who provided care that allowed the applicant to stay at home

A home can be transferred in limited cases to a child or a sibling. However, the transfer of real property has income, estate, and gift tax and other legal consequences, so you should have a qualified attorney review and complete the transaction.

Whole Life Insurance

Life insurance with a face amount (death benefit) of $1,500 or more is considered an available asset. Does this mean that life insurance needs to be cancelled to qualify for Medical Assistance? No.

Here's an example: The death benefit is $50,000, the cash value is $20,000, the amount at risk is $30,000. The life insurance may be kept in force in two ways. First, the cash value ($20,000) maybe made part of the CSRA. Second, the cash value may be borrowed from the policy and invested in an annuitized annuity and made an unavailable asset. To keep the policy in force, the spouse or the beneficiaries would need to pay loan interest.

Rental Property

Rental property is an exempt asset if it is the primary business of the applicant or spouse. Net income produced is used either as part of the applicant's share of nursing home costs or to provide income for the stay-at-home spouse. If it's not part of a business, it's a countable asset and can be made exempt or unavailable as part of an expanded CSRA through a waiver or an administrative hearing.

Pensions and IRAs

IRAs and pensions, minus any early withdrawal penalty, are an available asset. Under federal tax law, people who are retired and age 70½ or older must take a minimum distribution from IRAs and pensions. The distribution may become part of the Minimum Monthly Maintenance Needs Allowance. Pension accounts may be considered an unavailable or exempt asset if they are under a monthly distribution plan, but they will be subject to the income rules.

Gifts and Asset Transfers

An applicant for Medical Assistance is allowed to make unlimited transfers to his or her spouse and disabled children. These transfers do not create a period of disqualification and they are not subject to a look-back period. However, as we mentioned earlier, other transfers are subject to a look-back period—thirty-six months for transfers to individuals and sixty months for transfers to trusts. Any transfers that do not meet the test may result in a penalty period.

> **Look-Back Period:** A waiting period before a Medicaid application is filed or the applicant enters a nursing home.

There are special rules for giving away property or income. If the applicant and spouse give away less than a total of $500 in a month, there's generally no penalty. If the amount exceeds $500, there may be a penalty period.

Trusts

If an individual, his or her spouse, or anyone acting on the individual's behalf establishes a trust using at least some of the individual's funds, that trust may be considered available in determining eligibility for Medicaid.

No consideration is given to the purpose of the trust, the trustee's discretion in administering the trust, restrictions in the trust, exculpatory clauses, or restrictions on distributions. How a trust is treated depends to some extent on what type of trust it is—for example, whether it is revocable or irrevocable and any specific requirements and conditions.

This is how trusts are treated generally:

- Amounts actually paid to or for the benefit of the individual are treated as income to the individual.
- Amounts that could be paid to or for the benefit of the individual, but are not, are treated as available resources.
- Amounts that could be paid to or for the benefit of the individual, but are paid to someone else, are treated as transfers of assets for less than fair market value.
- Amounts that cannot, in any way, be paid to or for the benefit of the individual are also treated as transfers of assets for less than fair market value when contributed to the trust.
- Trusts established by others are not treated as being available.

In all of the above instances, the trust must provide that the state will receive any funds remaining in the trust when the individual dies, up to the amount of Medicaid benefits paid on behalf of the individual.

Certain trusts are not counted as being available to the individual. They include the following:

- Trusts established by a parent, grandparent, guardian, or court for the benefit of an individual who is disabled and under the age of 65, using the individual's own funds.
- Trusts established by a disabled individual, parent, grandparent, guardian, or court for the disabled individual, using the individual's own funds, where the trust is made up of

pooled funds and managed by a non-profit organization for the sole benefit of each individual included in the trust.
- Trusts composed only of pension, Social Security, and other income of the individual, in states that make individuals eligible for institutional care under a special income level, but do not cover institutional care for the medically needy.

In all of the above instances, the trust must provide that the state receives any funds, up to the amount of Medicaid benefits paid on behalf of the individual, remaining in the trust when the individual dies.

Health and Dental Insurance

An applicant for Medical Assistance must tell the county agency about any health or dental insurance policies that he or she has. If the local agency decides that a current policy will save money, MA will pay the premiums. Otherwise, any health or dental insurance would be unnecessary, since MA covers the following care:

- Physician services
- Home health care
- Most prescriptions
- Dental care
- Hospital and nursing home care
- Medical tests
- Physical therapy
- Eyeglasses
- Hearing aids
- Medical equipment

Long-Term Care Insurance

The median age of the United States population is at an all-time high. Adults over the age of 65 have surpassed the number of teenagers, and people in their 50s and 60s can expect to live longer than previous generations. As life expectancy continues to rise in the U.S., more and more Americans between the ages of 40 and 84, especially those in their mid 50s, are preparing for their golden years by purchasing long-term care insurance.

Long-term care refers to a wide range of medical, personal, and social services. You may need this type of care if you have a prolonged illness or disability. This care may include help with daily activities, as well as home health care, adult daycare, nursing home care, or care in a group living facility.

> **Long-Term Care Insurance:** Private insurance to cover the costs of long-term care (also called custodial care) for people with chronic health conditions and/or physical disabilities who are unable to care for themselves.

Long-term care insurance (LTCI) provides coverage for chronic illness and long-term disability not covered by Medicaid or Medicare. It generally covers the cost of nursing homes as well as certain agency services as visiting nurses, home health aides, and respite care. Your age, financial situation, and overall health will determine if this coverage makes sense for you.

According to some estimates, long-term care policies cost Americans, on average, $888 per year at age 50, $1,850 per year at age 65, and $5,880 per year at age 75. On a national average, nursing home care costs more than $51,000 a year. With costs rising with age, it is important for consumers to fully understand long-term care insurance and when it should be purchased to best prepare them for the future.

LTCI can help you preserve assets for family members if you don't want to spend down your savings to qualify for Medicaid. You can get individual coverage through most life insurance companies. You may also be able to get group coverage for yourself and possibly your parents through your employer or other associations.

The Health Insurance Portability and Accountability Act of 1996 (HIPAA) encouraged the use of long-term care insurance. It changed tax law for LTCI contracts that meet certain federal standards. In general, HIPAA treats certain qualified long-term care contracts the same as health insurance for tax purposes. The premiums for these contracts are deductible in whole or in part, the benefit payments are excluded from personal income, and the unreimbursed cost of qualified long-term care services are deductible as a medical expense. As with all insurance, you should carefully check the costs as well as the type and amount of coverage.

How much in benefits will the policy pay?

The benefit amount usually is a daily benefit ranging from $50 to $250 per day. You may choose a benefit period that is a specific number of days, months, or years. A maximum benefit period may range from one year to the remainder of your lifetime. It is important to ask the person selling the policy if the benefit amounts will increase with inflation and if that coverage increases your premium.

Are there exclusions?

Every policy has an exclusion section. Some states do not allow certain exclusions. Many long-term care policies exclude coverage for the following:

- Mental and nervous disorders or diseases (except organic brain disorders)
- Alcoholism and drug addiction
- Illnesses caused by an act of war
- Treatment already paid for by the government
- Attempted suicide or self inflicted injury

Considerations Before Buying LTCI

Whether you should buy long-term care insurance depends on your age and life expectancy, gender, family situation, health status, income, and assets.

- **Age and Life Expectancy:** The longer you live, the more likely it is that you will need long-term care. The younger you are when you buy the insurance, the lower your premiums will be.

- **Gender:** Women are more likely to need long-term care because they have longer life expectancies and often outlive their husbands.

- **Family Situation:** If you have a spouse or adult children, you may be more likely to receive care at home from family members. If family care is not available and you cannot care for yourself, paid care outside the home may be the only alternative. Different policies may cover different types of long-term care. It is important to buy a policy that will cover the type of care you expect to need and will be available in your area.

- **Health Status:** If chronic or debilitating health conditions run in your family, you could be at greater risk than another person of the same age and gender.

- **Income and Assets:** You may choose to buy a long-term care policy to protect assets you have accumulated. On the other hand, a long-term care policy is not a good choice if you have few assets or a limited income. Some experts recommend you spend no more than five % of your income on a long-term care policy.

GLOSSARY OF ESTATE PLANNING TERMS

A-B Trust: A common trust strategy created under a will to help maximize the unified credits and avoid estate taxes. By using it properly, a husband and wife can double shelter from estate taxes. This strategy is also available through a living trust.

Ademption: The removal of property from an estate by the owner after he or she has bequeathed it in a will.

Adjusted Gross Income (AGI): Your gross income reduced by certain adjustments.

Alternative Minimum Tax (AMT): A tax that you may pay instead of income tax if you have tax preference items or certain deductions allowed in determining regular taxable income.

Annuity: Investment that pays a fixed amount to a designated beneficiary for a specified number of years or for life.

Artificially Administered Sustenance: Medical treatment consisting of giving special nutritional formulas, fluids, and/or medications through tubes when a person cannot drink and/or eat, to sustain life although it cannot reverse the dying process.

Attorney-in-Fact: Person to whom you give the authority to conduct all affairs, make decisions, and carry out financial tasks on your behalf through a power of attorney.

Beneficiary: An individual who receives benefits from an estate or from assets that have been placed in trust.

Brain Death: Condition in which the entire brain has stopped functioning, so the person is dead according to established medical criteria, but kept breathing with a respirator.

Buy/Sell Agreement: The most common way to transfer ownership of a business when a partner dies: All partners in a business agree to purchase the interest of any partner who dies. These agreements are often funded by life insurance. Also known as a cross-purchase agreement.

Bypass Trust: A trust that is set up to bypass the surviving spouse's estate, thereby allowing full use of the personal federal estate tax exemption for both spouses. Also known as the Credit Shelter, Family Trust, B Trust, or Family Credit Shelter Trust.

C Corporation: A conventional corporation that pays tax directly to the IRS. It may be publicly owned or closely held. (C corporations are named after Subchapter C of the Internal Revenue Code.)

Capital Gains Tax: The income tax that must be paid when appreciated assets are sold at a profit or when a depreciated asset is sold at more than its book value.

Cardiopulmonary Resuscitation (CPR): A procedure used to restore breathing and a heartbeat, usually by applying pressure on the chest to keep blood flowing and sometimes by inserting a tube through the mouth or nose to get air into the lungs, connecting a ventilator to maintain breathing artificially, or using drugs or electrical shock to restart the heart.

Carrier: A private insurance company that contracts with the federal government to provide Medicare Part B coverage.

Charitable Remainder Trust (CRT): A gift made in trust to a qualified charity, an arrangement that regularly pays income from the assets to the donor or another beneficiary during the donor's lifetime and then passes the remaining assets to the designated charity.

Codicil: A legal change to a will, written and properly witnessed.

Conservator: A person appointed by the court to be legally responsible for managing the financial affairs of a person who's incompetent, playing a role similar to that of a guardian.

Conservatorship: The management of financial and personal affairs for a person who has been declared legally incompetent by a court, which appoints a conservator for that purpose.

Consumer Price Index (CPI): A federal measurement of inflation and deflation based on changes in the relative costs of goods and services for a typical consumer.

Contingent Beneficiary: An individual who is entitled to receive the benefits of an insurance policy if the primary beneficiary dies.

Contingent Trustee: A trustee whose appointment is dependent upon the original trustee's inability to act.

Credit Shelter Trust: A trust that reduces estate taxes by using the unified credits of both husband and wife and generates income for the surviving spouse. Also known as credit shelter family trust, family trust, bypass trust, B trust, family credit shelter trust, credit trust, and exemption trust.

Cross-Purchase Agreement: A contract that provides for an orderly transfer of ownership interests (stock or partnership interests) in a closely held business when an owner dies or becomes disabled. These agreements are often funded by life insurance. Also known as a buy/sell agreement.

Declarant: A person who makes a living will.

Declaration to Physicians: A form of living will, a document that specifies your wishes concerning health care treatment in the event that you are no longer able to make such decisions.

Devise: (as a noun) A bequest or gift in a will; (as a verb) to bequeath or give in a will.

Dialysis: Medical treatment used to remove waste products from the blood when the kidneys fail to work properly.

Disclaimer: A formal legal refusal by a person to accept property willed to him or her. The property then passes to the next person in the line of succession.

Do Not Resuscitate (DNR): A physician's order that, in the event of a sudden cardiac or respiratory arrest, no cardiopulmonary resuscitation (CPR) will be initiated, although all other medical and nursing care will continue.

Durable Power of Attorney: A written document by which a person designates another person to act on his or her behalf. It is not terminated by subsequent disability or incapacity of the principal.

Durable Power of Attorney for Health Care: A written document by which a person designates another person to act on his or her behalf to make health-care decisions if he or she is unable to do so.

Estate: All the assets owned by an individual at death, including home, real estate, bank accounts, securities, retirement plans, life insurance, etc.

Estate Liquidity: The extent to which an estate consists of cash or assets that can be easily converted to cash with little or no loss of value.

Estate Planning: Planning for management of assets during life and for orderly distribution of assets at death to heirs with the least possible delay and cost.

Estate Tax: A transfer tax imposed on the fair market value of property left at death; often called an inheritance tax or a death tax.

Executor: The person or institution named in a will to be responsible for the management of the assets and the ultimate transfer of the property; also commonly called a personal representative.

Fiscal Intermediary: A private insurance company that contracts with the federal government to provide Medicare Part A coverage.

Generation-Skipping Transfer: The passing of assets from the owner to his or her grandchildren, so that they are never in the possession of the owner's child or children.

Generation-Skipping Transfer Tax: A tax levied on assets that are transferred directly to grandchildren or lower generations.

Generation-Skipping Trust: A trust that allows assets to bypass a generation, so that grandchildren receive property directly from their grandparents, without it passing through their parents.

Gift: Any voluntary transfer of property or property interests to another without adequate consideration.

Gift Tax: A tax imposed on transfers of property by gift that exceeds the annual gift exclusion allowance, currently $13,000 per recipient per year.

Grantor: The person who sets up the trust, names the beneficiary and the trustee, and transfers the assets to the trust.

Guardian: A person who's legally responsible for managing the affairs and the care of a minor or a person who's incompetent; in some states a conservator plays a similar but more limited role.

Health-Care Proxy: A durable power of attorney for health care that appoints an agent to make decisions about medical treatment for the principal.

Heir: Person who inherits property when somebody dies.

Incapacitation: In general, the loss of mental competence; the inability to make decisions.

Individual Retirement Account (IRA): A financial construct that allows you to contribute to an interest-earning account for a specific purpose, originally retirement but now education as well.

Insurance Trust: Irrevocable Life Insurance Trust (ILIT). A type of irrevocable trust used to maintain a life insurance contract outside of an estate. Typically used to provide estate liquidity and/or to pay estate taxes and settlement costs.

Intestate: The state of dying without a will. Assets are then distributed through the probate process and according to the state's will for an individual.

Intestate Succession: The process for determining what will happen to the property and any minor children of a person who dies without a will (intestate).

Irrevocable Life Insurance Trust (ILIT): A trust that owns a life insurance policy, so that death benefit proceeds do not enter the estate and get taxed.

Irrevocable Trust: A trust that the trustor (grantor) cannot revoke or change.

Joint Tenancy: A form of ownership that provides for distribution of an interest at death to the other joint tenant(s). Joint tenant property is not transferable by will. Joint tenancy avoids probate.

Joint Tenants in Common: Owners of a shared asset, with the interest of any owner, upon death, becoming part of that person's estate.

GLOSSARY OF ESTATE PLANNING TERMS

Joint Tenants with Rights of Survivorship: Owners of a shared asset, with the interest of any owner, upon death, passing to the surviving co-owners.

Letter of Instructions: A memo that contains such information as the location of your will, the location of other vital documents, and any wishes for your funeral and burial.

Life Estate: A type of ownership that splits an asset into two parts: Full benefit during life for the owner or life tenant and the remaining interest passing directly to heirs (remaindermen).

Life-Sustaining Treatment: Medical treatment that allows life to continue, although it cannot cure or reverse the patient's condition.

Life Tenant: A person who makes a lifetime transfer of a property and retains all use of the property until death, at which time it passes to the remainderman.

Liquidity: The degree to which an asset can be converted into cash quickly and with little or no loss of value.

Living Probate: The court-supervised process of managing the assets of one who is incapacitated.

Living Trust: A trust that a person creates during his or her lifetime to manage assets and ultimately distribute them upon incapacity or after death.

Living Will: A document in which a person specifies the kind and extent of medical care he or she wants in the event that he or she becomes terminally ill and incapable of expressing his or her wishes.

Long-Term Care Insurance (LTCI): Private insurance to cover the high costs of long-term care (also called custodial care) for people with chronic health conditions and/or physical disabilities who are unable to care for themselves.

Look-Back Period: A waiting period before a Medicaid application is filed or the applicant enters a nursing home.

Marginal Tax: The tax imposed on an estate that is valued in excess of the unified credit exemption, when the value of lifetime gifts has been included.

Medicaid: A joint federal/state program that provides medical care to the needy; the federal government sets minimum standards that all state Medicaid programs must meet and then each state decides how much its programs will cover.

Medicare: A federal health insurance program primarily for people over age 65 who are receiving Social Security retirement benefits.

Medigap: Health insurance policies that supplement Medicare coverage.

Per Capita: Distribution that divides property equally among a group of named beneficiaries, regardless of their degree of kinship to the decedent. Example: A daughter, a grandson, an aunt, and a great nephew would all take equal shares if they were in the designated per capita class.

Per Stirpes: Distribution by line of descent, with any children of a deceased beneficiary splitting his or her share equally. ("Per stirpes" is a Latin term meaning by lineage.)

POD (Payable-on-Death) Accounts: A form of ownership to name a beneficiary on a bank or financial institution account, typically used in lieu of joint ownership. Also known as a Totten trust.

Pour-over Will: A will or a provision in a will that directs property to go to another legal entity, usually a trust.

Power of Attorney: Written authorization enabling a person to designate another person to act on his behalf.

Present Interest: The right to use a gift immediately.

Probate: The legal procedure to determine, if there's a will, whether the will is valid or, if there's no will, how the estate should be distributed.

Professional Corporation: A corporation composed exclusively of professional service providers, such as doctors, lawyers, accountants, architects, and others licensed to practice a "learned profession" or provide a service. It must file articles of incorporation with the state that meet the state's specific requirements.

Proxy (Health Care): A person designated in the living will to make health-care decisions for the declarant or principal when he or she can no longer do so.

Qualified Terminal (or Terminable) Interest Property (QTIP) Trust: A trust that uses the marital deduction at estate settlement and allows the grantor to determine to whom the trust assets will pass when the surviving spouse dies. The trust income must be paid to the surviving spouse and no person(s) can have the right to appoint the property to anyone other than the spouse during his or her life.

Reasonable Medical Practice: Practice that meets standards of care established by experience and provides reasonable expectation of benefit for a particular patient in a particular situation and in accordance with state law and organizational policy.

Remainderman: A person who has a future interest in a life estate or a trust.

Residuary Estate: What remains of an estate after all specific property bequests have been made.

Respirator/Ventilator: A machine that helps a person breathe or that substitutes for natural breathing when a person is unable to breathe because of illness or injury. Use of a respirator or a ventilator requires insertion of a tube through the nose or mouth or a tracheostomy (a surgical opening in the throat).

GLOSSARY OF ESTATE PLANNING TERMS

Retitle: To change legal ownership of an asset; the usual types of ownership are individual. Joint tenancy, or tenancy in common.

Roth IRA: Individual Retirement Arrangement created by the Taxpayer Relief Act of 1997 that differs from the traditional IRA in that contributions are not tax-deductible but withdrawals are tax-free.

S Corporation: A corporation that is taxed like a partnership. The corporation generally pays no tax; instead, all income and losses pass through directly to the stockholders, who pay taxes on their shares. (S corporations are named after Subchapter S of the Internal Revenue Code.)

Simplified Employee Pension (SEP): A simple, inexpensive pension plan designed for small business owners and self-employed individuals as an alternative to the 40l(k), profit sharing, and pension plans.

Special Needs Trust: A legal arrangement that allows a person to provide for a disabled loved one without interfering with government benefits.

Stepped-up Basis: For income tax purposes, the cost basis establishes the level over which capital gain on sale is assessed. Assets that are gifted to heirs during lifetime carry their original cost basis when transferred. Assets gifted at death receive stepped-up cost basis to the market value for estate tax purposes.

Survivorship Life Insurance: A life insurance policy that covers two people, usually spouses, and pays off only when the second person dies. Also known as joint, joint survivorship, two-life, or second-to-die.

Taxable Estate: The amount, after adjustments, of an estate value that is subject to federal estate tax.

Tenants in Common: A form of ownership that provides for collective ownership of undivided shares of an asset whereas the interest is transferable by will.

Terminal Condition: An incurable or irreversible condition in which any medical treatment will only prolong the dying process.

Testamentary Trust: A trust set up in a will and funded at death.

Testate: The state of dying with a valid will.

Testator/Testatrix: The man or woman who makes out a will and whose estate is to be distributed.

Title/Titling: Ownership of an asset, usually as individual owner, joint tenants, or tenants in common.

Totten Trust: A form of ownership to name a beneficiary on a bank or financial institution account, typically used in lieu of Joint ownership. The shared account belongs to the depositor until he or she dies, then passes to the designated beneficiary. Also known as a payable-on-death (or pay-on-death) or POD account, informal trust, or bank trust account.

Transfer Tax: A tax imposed when ownership of property passes from one person to another, as a gift or through a will.

Trust: A legal arrangement under which a person, or persons, or institution controls property for the benefit of the trust beneficiaries. The three parties to a trust include; the one who transfers property (trustor, grantor), the manager (trustee), and the beneficiaries. In the case of Living Trusts, the trustor, trustee and beneficiaries are usually the same individual(s).

Trustee: The individual or institution responsible for managing a trust.

Trustor: The individual who creates and transfers assets to a trust. Also known as a grantor.

Unified Credit: A credit the federal government gives each individual to use against federal estate taxes and excess gifts transferred during their lifetime.

Uniform Gifts to Minors Act (UGMA) Account: An account set up for a minor, with an adult designated as custodian of the property. The minor is the legal owner of the property, pays taxes on earnings generated by the property, and has an unrestricted right to use it upon reaching the age of majority (21 in Minnesota).

Uniform Transfers to Minors Act (UTMA) Account: The same as Uniform Gifts to Minors Act (UGMA) account.

Unlimited Marital Deduction: An individual can pass an entire estate regardless of size to a surviving spouse, without estate or gift taxes.

Vegetative State: A state in which a person is unable to talk or think or understand others. This condition, which can result from strokes and other diseases of the brain, is irreversible except in rare circumstances. A person in this state needs support in all aspects of care.

Wealth Replacement Trust: A trust set up to compensate heirs for a contribution to charity of assets that would otherwise have been included in the estate for the heirs.

Will: A legal document that is put into effect at the death of an individual. It serves as a list of instructions to the probate court on how estate assets should be distributed absent a joint owner or beneficiary designation.